Eco-Church

An Action Manual

Albert J. Fritsch, S.J.
with Angela Iadavaia-Cox

MEDIA LIBRARY
Department of Religious Education
225 Elm Street
Youngstown, Ohio 44503

Resource Publications, Inc.
San Jose, California

Editorial director: Kenneth Guentert
Managing editor: Kathi Drolet
Cover art: George Collopy and Huey Lee
Cover design: Huey Lee
Yurt drawing: Mark Spencer

©1992 by *Appalachia – Science in the Public Interest.* Published exclusively by Resource Publications, Inc. All rights reserved. For reprint permission, write:

Reprint Department
160 E. Virginia Street #290
San Jose, California 95112

Library of Congress Cataloging in Publication Data

Fritsch, Albert J.

    Eco-church : an action manual / Albert J. Fritsch with Angela Iadavaia-Cox.

        p. m.

    Includes bibliographical references.

    ISBN 0-89390-206-3: $14.95

    1. Human ecology—Religious aspects—Christianity. 2. Church management. I. Iadavaia-Cox, Angela, 1948- II. Title.

BT695.5.F76 1992   91-41762

96 96 94 93 92   5 4 3 2 1

Contents

Acknowledgments . . . . *vii*
Introduction 1

I. Getting Your Own House in Order: An Environmental Audit for Churches   7

Prerequisites for an Audit 9
Know Your Facility 9
Physical Space and Plant Operations 11
Building a New Structure 13
Conserving and Managing Waste 15
Land Use and Care 16
Food 17
Transportation 17
Recreation 18
Funding Resources 18
Group Discussion: Making Choices 21

II. Celebrating God's Earth   23

A Celebration Audit

Formal Church Worship 25
Ecumenical or Interfaith Celebrations 27
Outreach Programs 27

Ways To Celebrate

    Environmental Sabbath 29

    An Environmental Sabbath Celebration 30

An Earth Day Sermon 32

Confession Rites

    General Confession 35

    Prayer to Heal the Earth 35

Ecological Interfaith Cooperation

    The Assisi Declarations: A Call 36

    The Shakertown Pledge 40

Other Ways to Pray

    Guided Meditations 41

    Retreat Ideas 45

III. A Critical Look at Lifestyles   51

Lifestyle Change

    A Necessity for the Affluent 52

A Different Way of Living 54

How Much Do You Use?

    Food 57

    Housing 58

    Clothing and Personal Items 59

    Yard Care 60

    Travel 60

    Recreation 61

    Social Services 61

    Community Activities 62

What Happens To Your Time?

    Personal Time Budget 63

    Group Discussion: Personal Lifestyle 65

A Reflection on Lifestyles

    Quality of Life 66

    Charity—or Justice? 66

    Creating an Environment 67

IV. Healing the Earth: Options for Action   69

Mobilizing For Action

    Select a Cause 70

    Build a Base of Support 70

    Encourage Volunteers 70

    Spread the Word 71

Educating Your Membership

    Start a Library 72

    Sponsor a Discussion Group 72

    Host Environmental Awareness Events 73

    Utilize Church-affiliated Education Programs 75

Creating Sacred Spaces

    Home as Sacred Space 77

    Church as Quiet Space 78

    Outdoor Sacred Space 79

    Sacred Space in the Heart 79

Going Political

    Direct Lobbying 80

    Local, State and National Appointments 80

    Party Participation 80

    Political Office 81
    Testimony and Comments 81
    Voting Registration 81
    Oversight and Regulations 81
    Targeting Funds and Resources 82

Bringing Hope Through Alternatives
    Start a Community Garden 83
    Get Involved in Third World Issues 84
    Share Appropriate Technology 84
    Become a Peace Advocate 85

Environmental Resources

Educational Resources 91
Organizational Resources 95
Books of Interest 107
Solar-Heated Churches 113
Land and Rural Issues 117
Hunger Issues 119
Ecological Merits of a Simple Lifestyle 121

Acknowledgments

We would like to thank the many individuals who contributed to this book. We would especially like to thank Robyn Arnold for her coordination and layout of the material and John Freeman for his extensive help with the first draft of *Earthen Vessels,* the precursor of *Eco-Church*.

We acknowledge the numerous groups and individuals who allowed us to incorporate materials including: Roy Johnson, John Milton, Robert Sears, Sophie Jakowska, Jane Blewett, Mark Spencer, and the United Nation's Environmental Programme.

Introduction

Today, God's Earth is in deep trouble. Pollution deteriorates the planet's air, water and land. Oil spills blacken commercial waterways. The upper atmosphere ozone layer is depleted and hazardous waste contaminates municipal landfills and drinking water sources. Pesticides are used in ever-increasing quantities. The litany of woes is almost endless. Truly, there is an urgency to do something to save the Earth from destruction.

Many people no longer believe that the Earth can be saved (or at least they live as though they do not believe it). Yet ecologists affirm that the destructive processes can be reversed, even at this late date. While damage to the Earth is severe, some precious moments remain for the necessary changes.

Who is better suited to respond to this Earth-saving call than those who accept the role of co-creators of a New Heaven and a New Earth (Isa 51:16; 65:17; Rev 21:1)? Many church communities are committed to acting as leaven in the dough of a distracted and unbelieving world. These communities find power in their collective affirmation of what is good and just in a greedy and wasteful culture. They are powerful signs of God's witness among us.

The emphasis in this book is not individual environmental action, but rather the efforts of worshipping communities. These faith communities can support like-minded groups and create germinal environments. These new groups can, in turn, replicate themselves throughout the world.

Eco-Church is written for church leaders at both local and regional levels, i.e., pastors, assistants, social justice coordinators, teachers and educational administrators, Bible study leaders, youth program directors, organizers of church outreach, retreat and recreational programs, spiritual directors, those in resource conservation and stewardship efforts, as well as church administrators. Since the authors are Christian, much of the suggested material will be most relevant to Christian communities. However, *Eco-Church* attempts to be ecumenical in scope by drawing from the rich resources of the the Judeo-Christian tradition and to a lesser degree, from other religious traditions. Being faithful to ecological conservation principles means to incorporate and conserve Biblical and traditional spiritual materials. We cannot omit our traditional spirituality but must incorporate it into our quest to save the Earth.

As the church community is a leaven for action so *Eco-Church* is intended to be an agent for change. There is no step-by-step process that is perfect. The goal of this book is to assist you in creating a model church environment. In this model, members work together to:

- use interior and external space well;

- conserve energy and secure at least a portion of it from local or renewable energy sources;

- be model stewards of the resources used in office, transportation, and recreation activities;

- transform the physical plant so that it becomes a model and a symbol of what is possible for others to do;

- understand ecological principles, celebrate their application, and teach them to others in the larger community;

- do all of the above with the least amount of environmental impact.

Beyond this, *Eco-Church* encourages you to work enthusiastically and share resources with similar communities in an ecumenical and interfaith manner.

This book can be a resource for action in a variety of ways and in different circumstances. For example, one community may include those with a heightened environmental consciousness. This community may quickly transform their church into an environmental model. Another group may wish to linger and to spend time discussing the local or regional environmental problems that affect the entire community. The group might need to reach an environmental consensus before launching into building a model environment within the church community.

What we want to emphasize is that the major road block to action is not lack of knowledge, but rather the will to act. Part of the hesitancy, however, will be attributed to lack of assembled information. Therefore, a rather extensive (but not exhaustive) list of organizations is included in Organizational Resources at the end of the book. In point of fact, many church communities are reluctant to do something different from other local churches. The hope is that yours will be the first. Pray and strive to be a model of service to the greater community's Earth- saving undertaking.

But community should not stop with the local church. Religious environmental awareness and action can and must strengthen and ferment the entire world's Earth-saving commitment, a commitment often lost through fatigue and distraction. Through the Spirit, all forms of action are important and all add to the Earth-renewing process. God works in many ways; so will church communities. Thus the richness of environmental action is preserved in all its manifestations.

Eco-Church: An Action Manual

I. Getting Your Own House in Order: An Environmental Audit for Churches

There is a story of a fifth century monk who went out with his young companion to preach. They simply walked through the town. No word was spoken. The young man became impatient and asked when the preaching was to start. The older monk replied that their example of modesty and humility was an eloquent sermon.

Individual churches can develop or enhance the environmental consciousness of their particular locality or region. One important means of doing so is through example. This chapter will help you analyze your institutional lifestyle patterns. By knowing what areas need improvement, you can reexamine your goals, do some practical long-range planning, and begin to convert your facilities into an environmentally harmonious site.

We say *begin* for two reasons. For one, it is beyond the scope of this book to present a complete list of steps that will transform a church into a model environmental institution. Secondly, we understand full well that for many churches and their membership, change is a slow and sometimes difficult process. But we encourage you to start. Often taking one step eases the anxiety and uncertainty that builds up when considering the great many things that have to be done. The point is to do something, don't just talk about the problems and what *should* be done to solve them.

The audit covers a range of topics, from what kind of food is served at church functions to how space is used. Some adaptations of the audit will be required, based on the type of community you have, e.g., an intentional religious community, a church in an inner city or in a rural area, or a summer youth camp. You will need to keep the goals, aspirations, values, and specific activities of your group in mind.

Perhaps as a church leader, you can incorporate some suggestions easily. In that case, go through these pages, choose one or two areas and start! Once you feel comfortable, go on to another area and so on. On the other hand, if you need to gather the support of others, you must give careful attention to the auditing process. Some people find change particularly difficult. Others can feel threatened personally by the presumed implications of the audit. Still others want clarification. To move from a more complex to a simpler lifestyle usually requires education and reflection. The process you choose should allow time for growth. Confrontation can be resolved through discussion, reflection, and prayer. Let your shared faith and commitment to spiritual values guide you.

The community analysis needed for the audit will require regular meetings, say weekly or monthly. A reasonable period between meetings will give the participants time to reflect. Pertinent literature is helpful (as is humor) in maintaining the interest of the group. People do not enjoy dwelling on unpleasant things and exposing personal wastefulness is certainly not pleasant. A sympathetic and knowledgeable facilitator can make a difference. Select someone who can relate to people of different economic, political, and social backgrounds. A good facilitator can encourage simpler lifestyles without creating guilt or fraying tempers.

Prerequisites for an Audit

Anyone can conduct an audit but not everyone is willing to implement the changes. The following prerequisites will help to assure your group's ability and willingness to benefit from the audit. Here are the general prerequisites:

- a leadership team who understands what an audit implies and who is enthusiastic and willing to implement changes;

- a clear mission statement (including stewardship of resources, an awareness of the environmental considerations of the physical plant, e.g., a willingness of group to use all available environmentally benign resources such as solar energy);

- a community willing to participate in the process (not merely have others conduct the audit) and to implement the decisions resulting from the audit;

- a community willing to spread the word about their activities to other similar groups and to the public at large.

Know Your Facility

The more information that is gathered beforehand the better the audit can proceed. Think of this fact-collecting phase as the germ of a library of information about the physical plant. The material gathered could be cataloged and made available to interested parties. It is helpful to review the following before performing an audit:

See page 19 for groups that conduct audits.

- **grounds maps** showing vegetation and boundaries (for larger properties soil maps and topographic are most helpful);

- **floor plans** of all physical structures plus indications of current interior space use;

- **annual food, electricity and fuel expenditures** with as much detail as possible;

- **plans for future use** of buildings and grounds including building plans.

Depending on the extent of the audit, many larger institutions seek the assistance of experts to assemble data and analyze facilities and operations.

An Environmental Church Community

The Franciscan Community of Oldenburg, Indiana, a group of Roman Catholic religious women, decided in early 1990 to have an audit of their farm property made by the Kentucky-based *Resource Auditing Service*. The target facility included about 400 acres of land plus a large livestock barn, two brick residences, several farm buildings, and a large underground fruit cellar.

As a result of recommendations made at the conclusion of the audit, the community launched into a refurbishing campaign. The group cleaned up the grounds and renovated several of the buildings. The barn, with a 6000 square foot loft, was opened as a center for the Midwestern Bioregional Congress in the fall of 1991. Organic gardening began and better systems for the use of surface water and the waste stream were initiated. Then all the local Franciscans and nearby village community were invited to help with the clean up and renovation. On at least three separate occasions each year, both young and old people from the local religious and civic community participated. The Franciscans assembled documentation and publicized their efforts throughout the process. In fact, the successor of *Solar Lobby*, **Renew America**, designated their group in its 1991 environmental award-giving ceremony in Washington, DC.

One important part of the Oldenburg experience was the leadership team. This team gave complete support to the audit itself as well as enthusiastic approval of the resulting recommendations. While much can be said about inspiration from the ranks, the leadership role in a community is key. Without it, improvements do not occur.

A second important aspect is that the Oldenburg community reactivated its physical plant while also making it available to the general public as a center for environmental growth and awareness. The simultaneous physical development of the plant and the community's sharing of its environmental consciousness with others are important dimensions of an ideal environmental church community.

Physical Space and Plant Operations

Church buildings are outward manifestations of the inner graces of a spiritual community. Churches of the past were constructed by the people themselves out of native materials in ways that were energy- and resource-conserving for the day. But whether a church is modern or colonial, urban or rural, somber or light and airy, each can express the harmony that exists between God and creation. This harmony should be reflected in both the church's structure and its maintenance. Some questions to ask:

- ☐ Do administrators of your church try to save non-renewable resources, such as reducing use of electricity or gas?

- ☐ Do your administrators make the church building an example of good environmental practices?

- ☐ Does your church building reflect the true needs of your members?

- ☐ Is the decor of your church compatible with the needs of your neighbors on this planet as well as your community? What spirit does your decor convey? Is it excessively concerned with comfort or does it express a solidarity with the poor and needy?

- ☐ Is space efficiently scheduled? Is unused space shared with the larger community, such as civic or environmental groups for their purposes?

- ☐ Is the pioneer practice of contracting dwelling space in the winter and expanding in the summer utilized? Are unused rooms blocked off when not in use to save heating and cooling? Are extremes in temperature avoided?

- ☐ Are you aware of the impact of air circulation and humidity on the comfort level within your buildings? Have you ever considered vent fans as an alternative to summer heat? Do you utilize simple devices such as shades? Do you have trees planted for shade? If an air conditioning system is needed, is it energy efficient?

- ☐ Are all your appliances and purchases energy efficient?

- ☐ Are your lighting systems energy efficient and adequate? Have you considered the proper placement of light, energy saving bulbs, and the use of reflectors when appropriate?

- ☐ Are clean fuels such as natural gas used?

- ☐ Have alternative systems of energy, such as solar energy, been considered? What about wood, wind, and bio-gas (produced in the partial decomposition of manure)?

- ☐ Have conservation methods such as caulking, insulation and individualized thermostats been applied to all old buildings?

- ☐ Have you considered an active or passive solar heating unit, modular space heaters, a solar green house, or a photovoltaic system? Do you know what alternative energy sources apply to your community?

- ☐ Do you know how much water you use? Are conservation methods put into practice only when a drought alert is on?

- ☐ Do you insulate your hot water heater? Have you installed solar water heating systems?

- ☐ Do you use water-efficient toilets, low-flow aerators on sink faucets and low-flow shower heads in your facilities?

- ☐ Do you check for leaks in faucets and pipes regularly in all your facilities?

- ☐ Do you distinguish between uses that demand pure water for cooking and drinking and gray water or rain water for lawn watering, garden watering, and car washing?

- ☐ Has your energy use increased in the last five years? Do you know the reason?

- ☐ Have you considered doing a cost analysis of energy efficient systems and the pay-back for the conservation of resources?

- ☐ Have you checked indoor air for pollution from such sources as asbestos, radon, or faulty flues?

- ☐ Have you checked your water for lead content and impurities?

- ☐ Have you declared your church a smoke-free zone?

Building a New Structure

Consider the following before deciding to build:

- ☐ Is the building really necessary? Can you share a space with another church, thus eliminating the need to build?

  Often people want more space but fail to remember the advantages of better utilization of cramped and cozy existing space requiring less heat and maintenance.

- ☐ Does the need for more space require tearing down existing structures or can existing structures be modified for additional use?

  Remember that our culture pressures people to discard existing objects in order to encourage sales of new materials and services. Think twice before tearing down.

- ☐ Is the building's location consistent with the needs of the community?

  Often a building is erected without respect for its orientation and wind direction. Buildings should be near homes so that less travel is required to get to the desired activities.

- ☐ Are you considering alternative sources of energy such as passive or active solar space and/or water heating?

  Such design can save on future energy bills and be well worth careful consideration. If there is good wind potential, how about the use of this renewable resource?

- ☐ Is the new structure a multipurpose? Can the worship space be used for another activity?

- ☐ Can the space be designed for use on different occasions for education, recreation, worship, festive gatherings, and other community activities?

- ☐ Will the construction involve those whom the building serves?

  Some members may wish to participate in the planning stage. Others may want to help with clearing the land, final interior painting, or landscaping. The community will experience added meaning in buildings constructed through its own efforts.

☐ Can native materials be used?

In some cultures all building materials are close at hand (clay, wood, stone, sand, etc.). It saves transportation resources and it is more bioregionally appropriate to build with what can be obtained near at hand. Think seriously about this possibility.

☐ Is the building to last for the ages or just the present moment?

Much in our culture is temporary and considered "throwaway." Churches and other buildings constructed to last are symbols of a community intention to be rooted and attached to a given locality.

An Alternative Space

A Yurt contains aesthetically pleasing space. It is open and round and it has windows that let in air and light. The surroundings are enhanced by the presence of this type of structure. The building can be used as a meditation room, a prayer place, a chapel, or a hermitage where people can remove themselves for a special time of prayer. The cost may be as low as $1-2,000 depending on the size of the building and the amount of unpaid construction work utilized.

For information and/or plans write to **The Yurt Foundation,** *Bucks Harbor, Maine 04618.*

14 · Getting Your Own House in Order

Conserving and Managing Waste

Each year Americans throw away over 150 million tons of trash. Paper products, yard waste, aluminum containers, plastics, and glass are flooding landfills beyond capacity. "The golden rule for waste reduction," according to Arthur H. Purcell, author of *The Waste Watchers: A Citizen's Handbooks for Conserving Energy and Resources*, is: "make less and use less." Does your church follow these principles?

USE FEWER PRODUCTS

- ☐ Does your church purchase out of habit or necessity?
- ☐ Does it use permanentware or reusable plasticware at all functions?
- ☐ Does it conserve paper by using recycled paper and paper scraps and by returning junk mail with a notice to cease?
- ☐ Does it buy office, food and garden supplies in bulk in order to eliminate unnecessary packaging?
- ☐ Does it belong to co-ops whenever possible?
- ☐ Does it make a practice of buying products that can be reused, refilled or repaired?

USE PRODUCTS LONGER

- ☐ Does your church have a repair mentality, whether it be for a chair or a lawn mower?
- ☐ Does your church discard products only after their useful life is over?

REUSE DISCARDED PRODUCTS

- ☐ Does your church recycle by separating the garbage and collecting paper, glass, tin, bottles, and plastic?
- ☐ What percentage of garbage from your church is yard waste? Does your church have a compost center to use for fertilizer?
- ☐ Has your church thought of ways to increase the recyclable percentage of the garbage?
- ☐ Would your church sponsor a recycling program, if necessary?
- ☐ Does your church sponsor clothing, toy, or other collections for the less fortunate? Does your church participate in thrift shops or rummage sales?

USE SIMPLER PRODUCTS

☐ Does your church use products that are made from energy-efficient material, such as cellulose insulation?

☐ Does your church use alternative cleaners such as baking soda, vinegar or lemon juice?

☐ Has your church banned the use of styrofoam products?

DISPOSE WITH CARE

☐ Does your church avoid the use of toxic materials whenever possible? (Toxic waste includes everything from batteries and lawn chemicals to household cleaning supplies and paint.)

☐ Does your church dispose of toxic waste through safe means as designated by your local government (e.g., properly labeled and sealed paint solvent) rather than through public garbage collections?

☐ Has your church encouraged the local government to sponsor a toxic waste clean-up day? Would your church host such a day?

Land Use and Care

The way you care for the land shows your sensitivity to God's creation. Fifteen percent of American fertilizer goes to lawn care. Each year a large American lawn requires more fertilizer than is needed to feed a Third World family. A lawn is often a sign of an upper-middle class mentality. When you also consider that chemicals used to acquire rich and lush green grass — herbicides, other chemical pesticides, and commercial fertilizers — eventually find their way into the water supply, you have to accept responsibility on yet another level. Remember, lawns can become hooked on chemicals.

☐ Has your church analyzed its lawn and maintenance costs?

☐ Has your church switched from chemical pesticides, commercial fertilizers, and herbicides to natural alternatives?

☐ How much water is used to maintain the lawn? Does the church collect rainwater for watering lawns and gardens?

☐ Is the land for display or for use? Is some area set aside for contemplation?

☐ Is a large portion of land used as a wildlife sanctuary with trees, shrubs, and ground cover chosen to provide food and shelter for birds, small mammals and insects? Has the church considered wildflower patches?

These can be selected for easy, low-energy maintenance.

☐ Has the church converted some or all the land into edible landscaping for the benefit of the entire local secular community?

A rich variety of fruits, vegetables, nuts, berries and herbs can be as delightful to the eye as ornamental landscaping when arranged in an aesthetically pleasing manner.

Food

The foods you eat, drink, and serve express lifestyle values. Two important considerations are health and a recognition of the haves and the have-nots. Waste is the result of either uneaten food or overeating. How you grow, cook, and preserve your food as well as where you get your food is important.

☐ Does the food served in your facility promote good health?

☐ Does the selection encourage alternative protein other than meat? Does the choice offer white meat over red meat, less sugar, less saturated fat, and few harmful or unnecessary preservatives or additives?

☐ Do you ever give talks or sermons about food on occasions such as *World Food Day* (see page 74) or Thanksgiving?

☐ Is the amount of food served consistent with need so that waste is minimized?

☐ Does your church participate in a gleaning program to feed the local poor?

☐ Does your church use solar ovens and other economical means of cooking food?

☐ Do you favor home-grown or locally raised food over produce shipped long distance?

Transportation

A transportation audit includes a listing of the mode of travel used by your church, the distances traveled, the amount of time and mileage involved for work, recreation and business purposes.

- ☐ Is fuel economy a major criterion in selecting church vehicles?
- ☐ Do your church parking lots contain bike racks?
- ☐ Are your members encouraged to carpool and walk whenever possible?
- ☐ Are important church events chosen with consideration of travel? Are buses and vans used instead of private cars? Are places closer to home chosen whenever possible?
- ☐ Does your church encourage legislation that supports public transportation?
- ☐ Does your church maintain its vehicles to minimize pollution?

Recreation

Whether it be an annual fundraiser or an end of the year celebration, what determines your selection of activity? Too often a community can hide behind traditions that are based on habit and lack of vision. Let re-creation come back into recreation.

- ☐ Does your church select activities with resource conservation in mind, such as taking church trips closer to home?
- ☐ Do any of your church functions celebrate the Earth? Do you educate your members about Earth's needs on days such as Earth Day, Arbor Day or World Food Day?
- ☐ Do your events foster good will? Do they encourage members to come together, to learn about other cultures?
- ☐ Does your church sponsor joint neighborhood church activities?

Funding Resources

If you reflect on religious retirement communities of the turn-of-the-century that were often rural and furnished their own food and fuel, and then compare their modern counterparts, you can see but one example of the need for financial investments by church establishments. Escalating prices and expenses mean higher earnings but there is a need to demonstrate conservation consciousness toward money. Attempting to do more with less has a spiritual foundation. It foreshadows future eco-based economies.

"Anytime money is saved through efficient use of resources," says conservationist John Freeman, *"the funds saved should be used for some activity having a low ratio of energy consumption to dollars spent."*

☐ Is your church's manner of spending and acquiring money Earth-compatible?

☐ Are investments made only with groups that are environmentally and socially ethical?

☐ Does your church use money saved through conservation wisely? Do you give the money to enhance the lives of people who do not have the basics of life?

Each church can set up its own charity or development mini-fund to be used for better housing, food production or energy conservation for needy people. In this way, the saving will be a true savings and also will help others.

Total Resource Audits

A resource audit is a comprehensive examination of grounds, physical plant, and energy use of a given organization or unit. The audit both describes the current condition of the facility and suggests improvements.

The reports from the audit will include alternative grounds uses, possible solar and renewable energy utilization, and suggestions for environmental programs, such as waste utilization and constructing safe indoor environments. For a total resource audit, write: Resource Auditing Service, P.O. Box 298, Livingston, KY 40445.

If you would like a professional team who specializes in Church Energy audits, contact: Interfaith Coalition On Energy, P.O. Box 26577, Philadelphia, PA 19141. (215-635-1122).

The UCLA Environmental Study Group released a report entitled *In Our Own Backyard: Environmental Issues at UCLA, Proposals For Change and the Institution's Potential as a Model*. For a copy of the report, send a check for $30.00 payable to the UC Regents, to: UCLA Graduate School of Architecture and Urban Planning, 405 Hilgard Avenue, Los Angeles, CA 90024-1467, Attention: Urban Planning Publications Director.

For information on alternative energy, conservation, nutrition, gardening and other topics of interest see *Organizational Resources*, and *Books of Interest* at the end of this book. For a list of solar-heated churches by denomination, see *Solar-heated Churches* in the same place.

Act Locally, Think Globally

Our pioneer tradition called for self-reliance. This spirit is not just North American, but shared by people of many lands and many ages. Self-reliance is a philosophy worth reconsidering.

Complex societies lose their self-reliance when they depend on others for the basics of life. Clothing excepted, the essentials (food, water, fuel, and building materials) are quite bulky. If each person in North America uses about 50 gallons of water a day, individual annual consumption weighs 70 tons. Food for that person amounts to one ton, and fuel another ton. Self-reliance reduces the need to transport these heavy items from other parts of the world. You can collect water and solar energy, grow food in backyards or on nearby farms, and use local building materials. Besides saving resources expended through transportation, you have more control over your basic resources.

This kind of self-reliance is a community condition, not an individual act of independence. Individuals within a community cannot survive without the physical assistance of others. Spiritually, people also need others. We need ideas, enthusiasm, new ways of seeing the world. Self-reliance is a quality of physical well-being that creates an atmosphere in which the spirit can flourish.

Local self-reliance also implies a sharing among equals, a reciprocity, and a dignified atmosphere that enables people to take control of their lives. Great disparities among members should not exist. A self-reliant community does not seek to be a benevolent autocracy, but to devise a democratic process in which the powerless gain respect. It seeks a process that allows all to become self-reliant. In optimal communities, members provide health, education, and other essential services. Self-reliant communities also have the means to assist other communities when needs arise — at times of earthquake, famine, or flood. Sensitivity and charity enhance and purify self-reliance.

The local community is like a living body. It has nerves, arteries, a heart, a brain, and an organic spirit. Our communications system is the nerves; our health services, our markets and transportation network, the arteries; our educational system the brain. A healthy body works as one unit. But the spirit is not contained locally. Though locally oriented, the self-reliant think globally. They recognize that everyone is a part of a spiritually interdependent Earth family. Their compassion and love extends beyond the confines of geography.

- What are the ways your local church can build a sense of community through locally-controlled cooperatives?

Group Discussion

The following questions are best considered by a church environment committee or by an administrative advisory or decision-making group.

Making Choices

- Is your church willing to commit itself at this time to a complete environmental audit? Are the church members and donors willing?

- Will you need to develop an audit preparation process for the church members? What steps should you take to see the process through to completion?

- What will you do with information gathered from the audit? What options will you consider that consume fewer resources?

- How can others benefit from your own resourcefulness — through example, through actual savings, through solidarity, or through resource sharing?

- Can your church's applications be replicated by other churches and other institutions? What ways can you use to publicize results of your church's conservation measures?

- Where does your church group invest its surplus? In the local bank? In community funds? In cooperatives? In investments outside the community? Extend the discussion to your church finance committee.

- Is your church willing to sacrifice some of its own complacency and realize the need to work together to save the Earth? Is God moving you to work with other believers to renew the environment?

- Are you called to encourage the have-nots of this world to take what is rightly theirs in justice?

- Is your servanthood role one that offers encouragement rather than one that merely gives?

II. Celebrating God's Earth

Worship and celebration can reflect the bond that exists between the Earth and its people. This bond is intricate and has many different facets, all of which need to be nurtured. When believers come face to face with God's presence manifested in the Earth and its creatures, they can experience mystery, self-renewal, and gratitude. By witnessing the Earth's magnificence and its potency, its intricate systems that defy limited and finite human schemes, the worshipping community can regain a sense of wholeness as well as humility. Worship and celebration must foster this experience.

Similarly, when believers confront their finite work of renewing and protecting the Earth, they can feel a certain powerlessness before God. While this may be disconcerting, they can also experience consolation in the knowledge that their work is good, and that all creation is good. Their environmental work is worth celebrating; faith affirms that meaningful things can be done to save the Earth. Celebration gives a sense of hope back to creative workers and reminds them that all is not lost. Through celebration they can be energized and become all the more enthusiastic. This contagious enthusiasm can be caught by others who, in turn, are moved both to work and to enter into celebration. Gradually, through celebration, more community support will be given to Earth-renewal.

Worship and celebrations also act as correctives. The goodness of the Earth can easily be exploited by those who do not truly value its inherent worth. Worship can bring people down to Earth. It can remind believers that servants need not be "dominating lords," because those effecting change exercise a power that comes from Another.

Celebration is part of harmonious living and a balanced human environment. No religious tradition is without its celebrations. Virtually all faith traditions have a relationship with the Earth and thus include the Earth in its celebrations. For those of the Judeo-Christian tradition, the Earth is understood to be good and to be created by an all good God (Gen 1). The mountains, seas, and all creatures rejoice (Ps 148). Earth, the reflection of its Maker, is described as rejoicing because the Creator rejoices in it.

Having said this, many worshipping communities within the Judeo-Christian tradition have lost their "Earth-celebrational" sense. That is a story in itself. In this chapter we simply want to let you know some ways you can incorporate an awareness of the Earth into your community's worship and celebrations.

In order to incorporate the Earth in your celebration, there is little need to look to other cultures. The Judeo-Christian tradition is filled with celebration, especially in the more liturgically-oriented communities. Besides formal worship, there are revivals, picnics, bazaars, pot lucks, outings, homecomings, etc. Both you as a leader and your community need to understand and affirm the "Earth-dimensions" of your own rituals. You can re-introduce the lost Earth-dimensions. It is not necessary to insert esoteric spiritual traditions into your worship practice. Re-introduction and adaptation of your traditional practices are far better ways to strengthen the Earth-dimensions of your worship than the use of unfamiliar rituals. For example, you can use seasonal vegetation as decoration in your worship space.

Celebration may be of two types: a passive celebration of the created Earth — looking out, listening, and seeing what God has done; and an active celebration of the work done to save, renew and re-create the Earth — a celebration of the community's sacred work. The former is Earth-centered; the latter is human-centered. The first celebrates God's gift to us and the second the use of that gift by God's co-creators.

A Celebration Audit

Often churches profess concern about the Earth or about environmental activities, but omit the integration of celebration into these activities. A church or congregational audit is in order. Note that the celebration or worship may be formal or informal. Some groups are less liturgical than others. Consider only those questions related to your church worship practice.

Church celebrations can be divided into three areas: formal worship, ecumenical or interfaith formal worship, and informal worship or activities by specific groups within a church, (e.g., a retreat, a social activity like a picnic). Individual religious activities at home or at other settings can also be considered church celebrations, especially when the church leadership offers help and information about individual forms of religious activities.

Formal Church Worship

One of the best ways of celebrating in community is within the formal church worship time. A number of opportunities exist where environmental themes can be integrated into worship.

- ☐ Are environmentally-related homilies or sermons given in your church? Is there special mention of Earth celebrations within church homilies, church bulletins, flyers, or notices?

- ☐ Are any of the following days celebrated by your faith community? (The exact date is generally not essential.)

World Food Day or **Week**	(October 16, or the week of)
St. Francis Feast Day	(October 4th)
Earth Day	(in April about the 20th)
World Environment Day	(on or about June 5th)
Environmental Sabbath	(the first weekend in June)
Arbor Day	(date varies by state)
Fast for World Harvest	(Thursday before Thanksgiving)

- ☐ Are days that celebrate the Earth mentioned merely in passing, such as a special petition or prayer or is the ceremony significant enough to be understood by the participants? Are special songs, prayers, responses, skits, processions, or blessings ever included in worship?

Celebrating God's Earth · 25

- ☐ Are traditional petitioning events, such as Rogation Days (three days of solemn prayer for the crops preceding Ascension Thursday), still celebrated in your church?

    Note that rural celebrations have lost favor due to urbanization of church communities.

- ☐ Are there special prayers for growing seasons or crops in your community? Do you ever pray for rain or protection from natural calamities such as flood or tornadoes?

- ☐ Are there special blessings for the fields (e.g. an Easter Water blessing for the gardens in the parish)?

- ☐ Do you process out of doors to remind your community of the beauty of creation? Do you include hymns related to our dependence on God as Provider of the bounty of the Earth?

- ☐ How well are Thanksgiving or harvest celebrations conducted? Does the church have special sharing with the needy at the time of the national Thanksgiving?

- ☐ Is New Year's Day a time of new resolution and peacemaking? Is there an effort to consider making peace with the Earth?

- ☐ Are there any special Confession Rites which include questions or examination of your community's relationship with the Earth and its resources?

- ☐ Do your church buildings have signs or banners with natural colors and environmental themes? Are these appropriately used with the changing seasons?

- ☐ Does your church building have seasonal decorations at its entrance (e.g., autumn leaves in fall, budding willow branches in spring)?

Ecumenical or Interfaith Celebrations

The shared celebration of worshipping communities in the same locality is very important. Coming together as a combined community committed to saving the planet symbolizes the participatory character of environmental action.

- ☐ Does your church community make an effort to celebrate common days (e.g., Thanksgiving) with nearby church communities?

- [ ] Do any of these joint celebrations contain environmental or Earth-centered themes? Is there a special day each year when an interfaith (Jewish-Christian) activity occurs or is planned?

- [ ] Are there any efforts to assemble hymns, rituals, prayers, or other worship materials that can be shared jointly with other groups? Has your liturgical or worship committee ever consider starting such a file or library of materials?

- [ ] Are your church premises ever used by other environmental groups, religious or secular?

Outreach Programs

Most worshipping communities see a great need to go beyond their local bioregions and affect the lives of people elsewhere. This more evangelistic approach may be extended to environmental concerns.

- [ ] Does your church community sponsor retreats for adult, youth, special groups? Are any of these held in a natural setting? Are environmental themes included in the retreat content?

- [ ] Does your church sponsor volunteer environmental work programs, such as beautifying roads or rebuilding nature trails? Do volunteers share experiences with others in the church?

- [ ] Are there alternative vacation programs available for your church community, e.g., where individuals or groups will run environmental retreats for inner city youth?

- [ ] Is there any interest in spiritual formation programs at your church? Do these include Earth-celebration themes?

- [ ] Does your church ever go to sites that are environmentally damaged and pray for the healing of the Earth?

- [ ] Do you have experienced people within the church or locality who could be environmental resource people? Have any come forward to volunteer time for church outreach programs?

- [ ] Is there a staff person who collects and distributes materials for home or family religious celebrations, e.g., Advent candle lighting ceremonies, Lenten home devotions, or general educational materials? Are Earth or environmental themes suggested for such practices, such as use of native and seasonal decoration rather than purchasing from distant commercial operations?

- [ ] Are there times set apart for blessings of homes, animals, vehicles, gardens, or grounds? Is this done on a one-by-one basis, or is there a congregational program for such exercises? Are there church community gardens? Do the gardens have special ceremonies for the beginning or close of the season?

- [ ] Are there environmental or social justice materials included in the parish library?

Note: An audit can be a checklist of suggestions as well as a statement of what is accomplished. Few church groups make affirmative answers to the questions asked. The questions may stimulate interest in initiating programs. The audit is intended to be a catalyst.

Ways to Celebrate

Environmental Sabbath

The United Nations Environmental Program asks all congregations to set aside a special day to address our responsibility to care for the Earth. Called the *Environmental Sabbath* or *Earth Rest Day*, this day should coincide or fall near World Environment Day on June 5. A complete folio on Environmental Sabbath offers guidelines for prayer and song. The following material is from the 1990 folio. The theme for 1991 is Our Children — Their Earth. All of the main religions are represented. Write: United Nations Environment Programme, Two U.N. Plaza, New York, New York 10017.

Creation of a New Day

Imagine an infant nursing at his mother's breast. His mother grows weak. But the infant continues suckling, allowing her no rest, until she is dead.

Then the infant must die, too.

Humankind is such an infant...

We at the United Nations Environment Programme are sending out a call to the spiritual leaders of this earth to help STOP the sacrilege. Help us reach the center of change: the human heart.

Lead your congregations in a day of prayer, reflection, reverence and commitment to action. The Environmental Sabbath.

On this day, we restore the Earth.

We must remember that we are not God's only child. If we are indeed the steward of divine purpose, we must act on earth as angels of mercy, not angels of destruction. We must consecrate the great accumulated wisdom of Science — so often the servant of greed — to the service of Life.

We must find within ourselves the spiritual center which is observant, gentle, patient and desirous of harmony.

We must say, do and be everything possible to realize the goal of the Environmental Sabbath: an ecological society, one in which people live responsibly and co-operatively within and with the rest of Creation. We cannot let our mother die. We must love and replenish her.

There is much to be done. It can and should begin in the shrines of the spirit, in the mosques, temples, and churches of the world. Help us create a new future. Help us create a new day. The Environmental Sabbath.

On this day we restore the Earth!

An Environmental Sabbath Celebration

- Decorate your sanctuary with photographs of Earth images.

- Invite guest speakers or "representatives" from other species, i.e., plants and animals.

- Play appropriate music such as *Earth Mass* by Paul Winter (*Windham Hill* label), the hymns "For the Beauty of the Earth," "Sing Praise to God Who Reigns Above," or "Heaven and Earth and Sea and Air."

- Say prayers for the healing of the Earth.

SERMON IDEAS

- Describe the environmental crisis, or some aspect of it such as the destruction of the rain forests. Use scientific data. Highlight the urgency.

- Speak of the essential earth-human relationship, what it is and what the responsibility of humankind is.

- Announce to the congregation that this is a special day: a day on which we make our commitment to restore Earth.

- Point to various sources of inspiration such as role models like Albert Schweitzer or St. Francis.

- Draw from the holy words of your religion.

- Mention some initiatives at the global and local levels to show that the work is going forward.

- Direct the congregation toward action, using examples from personal experience, e.g. questions that helped you discover your own environmental attitude or helped you reflect on the meaning of the Earth. Tell a story about how God touched your life through the wonders of creation. Challenge the congregation to do the same.

ENCOURAGE YOUR CONGREGATION TO ACT

- To follow environmental stories in the media.
- To plant trees.
- To fight pollution.
- To write to elected representatives.
- To join a local environmental group.
- To enjoy nature and be kind to living things.
- To recycle cans, bottles, and paper whenever possible.
- To try not to waste.
- To cut down on energy consumption.

A Call to Prayer

We who have lost our sense and our senses — our touch, our smell, our vision of who we are; we who frantically force and press all things, without rest for body or spirit, hurting our earth and injuring ourselves: we call a halt.

We want a rest. We need to rest and allow the earth to rest. We need to reflect and to rediscover the mystery that lives in us, that is the ground of every unique expression of life, the source of the fascination that calls all things to communion.

We declare a Sabbath, a space of quiet: for simply being and letting be; for recovering the great, forgotten truths; for learning to live again.

An Earth Day Sermon

On April 22, 1970 thousands of schools and millions of individuals participated in activities celebrating our Earth and its natural resources and learning about environmental problems. This is a sermon that was distributed to churches across the nation for Earth Day 2 in 1990.

If this is truly a day the Lord has made, let us rejoice and see it as an opportunity. Being people on a faith journey, we must plant our feet firmly on Earth and recognize obstacles which block us here and now. We see magnificent mountains, sparkling lakes, whispering forests, singing birds, verdant wilderness, silent deserts. But we must also recognize the bulldozers and clearcutting equipment that scar magnificent mountains; the water and air pollutants that kill lakes and their fish life; the rain forests and the wildlife that are disappearing at a phenomenal rate; our deserts that roar with the sound of recreational vehicles and are marred by hazardous waste burial sites. Terrestrial life as we know it is changing — and it is not all for the better.

While reasons to celebrate our planet abound, stronger reasons to be sober and thoughtful exist. The last two decades have witnessed some environmental progress, but the Earth's life is being threatened in an ever-increasing degree.

Today, we must transform celebration into thoughtful action. The needs are urgent, the manner in which we participate is important, and the faith commitment to such actions is profound. We are called to:

> share the limited resources of the Earth in a proper manner with the less fortunate:

> exert an effort at actually cleaning up the pollution that afflicts our Earth;

> and profess a constant faith that our Earth can be renewed and not destroyed by the greedy and thoughtless.

The early believers shared all things in common and divided everything on the basis of each one's need. Today, billions of human beings cry out for the basic needs of food, clothing, shelter, education, and health. Plants and animals are threatened in ever increasing numbers with possible extinction. Poor people and poor Earth, a community of earthly suffering exists. We need to imitate the early Christian community's faith. We, too, are called to redistribute world resources according to need so that those who over-consume are made more healthy by consuming less, and those who are destitute will have the basics of human life. In doing so *all* will improve the quality of their lives.

The need to act today is the same as with the early community. However, the community has grown from a small one to a global one. As Tolstoy said in the same spirit of the Early Western and Eastern Church writers: "All have a right to the fruits of the earth as they have to the rays of the sun." The taking control of world resources so that some can spend them extravagantly is a perversity of justice. We, the Lord's ministers on Earth, must be willing to confront the injustice just as Jesus did with those who turned the Lord's House, meant for all the people, into a haven for the money-changers.

Saving the Earth takes effort and involves sacrifices. Praising God's goodness in creation involves participating in environmental actions to help save the Earth. Ways of acting are many, depending on our talents and inclinations. When we speak up for clean water, teach young and old to be caring, strive to convert some part of military defense to earth or eco-defense, or pray for the Earth and its advocates, we are making that effort. Knowing what action is most effective for us involves serious reflection and prayer.

Some of us hesitate to become environmentally involved. We are like Thomas standing back and refusing to accept the Lord of the Earth's plea for us to help. We insist on more proof that the Earth is wounded, or say that we are not properly educated, or are too inexperienced. If we cannot see the scars on the Earth or probe the decimated rain forest of the Amazon, then we will not believe the Earth is in trouble. When we can't travel or get this first-hand information, we excuse ourselves and leave Earth-healing to others.

Shouldn't we choose life which beckons us to help in the noble task of re-creation? Can we honor the Earth by consuming only what is needed, or continue to dishonor her by squandering and misusing the good gifts with which we are blessed? Do we treat the Earth with familial respect and teach others to do the same? Are we willing to do more than listen to brave words or good music?

Make this a day of rededication to a goal of sharing resources, becoming environmentally active, and professing our faith in the goodness of the Earth itself.

Father Albert Fritsch
April 22, 1990

Song of the Sun

O most high, almighty, good Lord God,
to You belong praise, glory, honor, and all blessing!

Praised be my Lord God with all creatures;
and especially our brother the sun,
which brings us the day, and the light;
fair is he, and shining with a very great splendor:
O Lord, he signifies You to us!

Praised be my Lord for our sister the moon,
and for the stars,
which God has set clear and lovely in heaven.

Praised be my Lord for our brother the wind,
and for air and cloud, calms and all weather,
by which You uphold in life all creatures.

Praised be my Lord for our sister water,
which is very serviceable to us,
and humble, and precious, and clean.

Praised be my Lord for our brother fire,
through which You give us light in the darkness:
and he is bright, and pleasant, and very mighty, and strong.

Praised be my Lord for our mother the Earth,
which sustains us and keeps us,
and yields divers fruits, and flowers of many colors, and grass.

Praised be my Lord for all those who pardon
one another for God's love's sake,
and who endure weakness and tribulation;
blessed are they who peaceably shall endure,
for You, O most High, shall give them a crown!

Praised be my Lord for our sister,
the death of the body, from which no one escapes.
Woe to him who dies in mortal sin!
Blessed are they who are found walking by Thy most holy will,
for the second death shall have no power to do them harm.

Praise you, and bless you the Lord,
and give thanks to God, and serve God with great humility.

St. Francis
1182-1226

Confession Rites

The social sin of environmental damage needs to be understood properly. The Church as one body must seek God's forgiveness for the damage done to the Earth. This can be done at such soul-searching times as Lent, Advent, or Yom Kippur. Worship might include an opening song, a prayer, an examination of faults, a scriptural reading, sermon, a prayer of forgiveness, a dance or closing song, and a blessing.

General Confession

For the times we have failed to think of the harm done to air, water, and land,
Lord have mercy.

For failing to conserve energy,
Lord have mercy.

For ignoring our civic responsibilities,
Lord have mercy.

For allowing ourselves to be saturated by the allurements of a consuming culture,
Lord have mercy.

For not living more simply so others could simply live,
Lord have mercy.

For failing to ask for forgiveness from God
and creatures,
Lord have mercy.

For not being thankful for the many gifts God has given,
Lord have mercy.

Prayer to Heal the Earth

*This is why the land is in mourning,
and all who live on it pine away,
even the wild animals and the birds of heaven;
the fish of the sea themselves are perishing.* (Hosea 4:3)

If land mourns then it has feeling. The Bible has many references to the suffering of mountains and land itself. Some believers go to the site of a massacre or battle or surface mine site and say prayers of reparation. They seek God's forgiveness for the wrongdoing that has occurred in that location.

People who participate in this outreach program testify that the barren lands and fractured communities have been healed. Communities are known to live in greater accord and the Earth again is fruitful.

For additional information on praying over environmentally wounded communities write: Robert Sears, S.J., Jesuit House, 5554 Woodlawn Ave., Chicago, IL 60637 or Mary Klimowski, 2815 West 98th Place, Evergreen Park, IL 60642.

Ecological Interfaith Cooperation

All participants must accept the idea that people need not share the same theological understanding *in toto* to work together to save the Earth. A question of what is a "believer" may arise, but drawing up credal formulae or documents prior to cooperative environmental activities may be a dead end. Unnecessary theologizing can leave the Earth to suffer by itself. It may be advantageous to call a moratorium on theological discussion, lest it become distracting and divisive.

Instead, every effort must be made to highlight the similarities of beliefs. Research the ways various belief systems preserve and improve the Earth. Discover what ways are held in common by these different systems.

All believers have gifts that are valuable for the cooperative enterprise of renewing the damaged Earth. All are challenged to find within their traditions those qualities that will contribute to this goal. All can respect and promote the contributions of other believers.

The Assisi Declarations: A Call

In September 1986, the World Wide Fund for Nature (WWF) celebrated its 25th anniversary by bringing together five of the major ethical systems of the world — Buddhist, Christian, Hindu, Jewish, and Moslem leaders. All were asked to declare how their faith lead them to care for nature. The following are excerpts from their declarations. For the complete statements, write: The Environmental Sabbath, United Nations Programme, Two U.N. Plaza, New York, NY, 10017.

FROM THE BUDDHIST DECLARATION:

Buddhism is a religion of love, understanding and compassion and is committed toward the ideal of non-violence. As such it also attaches great importance towards wildlife and the protection of the environment on which every being in this world depends for survival...

These teachings lead us to the following words by His Holiness the Dalai Lama:

"As we all know, disregard for the Natural Inheritance of human beings has brought about the danger that now threatens the peace of the world as well as the chance of endangered species to live.

Such destruction of the environment and the life depending upon it, is a result of ignorance, greed and disregard for the richness of all living things. This disregard is gaining great influence. If peace does not become a reality in the world, and if the destruction of the environment continues as it does today, there is no doubt that future generations will inherit a dead world..."

The Venerable Lungrig Nomgayal

FROM THE CHRISTIAN DECLARATION:

...Because of the responsibilities which flow from his dual citizenship, man's dominion cannot be understood as license to abuse, spoil, squander or destroy what God has made to manifest his glory. Then dominion cannot be anything else than a stewardship in symbiosis with all creatures. On the other hand, his self-mastery in symbiosis with creation must manifest the Lord's exclusive and absolute dominion over everything, over man and over his stewardship...

...in the name of Christ who will come to judge the living and the dead, Christians repudiate:

1. All forms of human activity — wars, discrimination, and destruction of cultures — which do not respect the authentic interests of the human race, in accordance with God's will and design, and do not enable men as individuals and as members of society to pursue and fulfill their total vocation within the harmony of the universe.

2. All ill-considered exploitation of nature which risks to destroy it and, in turn, to make man the victim of degradation.

Father Lanfranco Serrini

FROM THE HINDU DECLARATION:

Not only in the Vedas, but in later scriptures such as the Upanishads, the Puranas and subsequent texts, the Hindu viewpoint on nature has been clearly enunciated. It is permeated by a reverence for life, and an awareness that the great forces of nature — the earth, the sky, the air, the water and fire — as well as various orders of life including plants and trees, forests and animals, are all bound to each other within the great rhythms of nature. The divine is not exterior to creation, but expresses itself through natural phenomena...

Let us declare our determination to halt the present slide towards destruction, to rediscover the ancient tradition of reverence for all life and, even at this late hour, to reverse the suicidal course upon which we have embarked. Let us recall the ancient Hindu dictum: "The Earth is our mother, and we are all her children."

Dr. Karen Singh

FROM THE JEWISH DECLARATION:

The highest form of obedience to God's commandments is to do them not in mere acceptance but in the nature of union with Him. In such a joyous encounter between man and God, the very rightness of the world is affirmed.

The encounter of God and man in nature is thus conceived in Judaism as a seamless web with man as the leader, and custodian, of the natural world. Even in the many centuries when Jews were most involved in their own immediate dangers and destiny, this universalist concern has never withered...Now, when the whole world is in peril, when the environment is in danger of being poisoned, and various species, both plant and animal, are becoming extinct, it is our Jewish responsibility to put the defense of the whole of nature at the very center of our concern...Man was given dominion over nature, but he was commanded to behave towards the rest of creation with justice and compassion. Man lives, always, in tension between his power and the limits set by conscience.

Our ancestor Abraham inherited his passion for nature from Adam. The later rabbis never forgot it. Some 20 centuries ago they told the story of two men who were out on the water in a rowboat. Suddenly, one of them started to saw under his feet. He maintained that it was his feet. He maintained that it was his right to do whatever he wished with the place which belonged to him. The other answered him that they were in the rowboat together—the hole that he was making would sink both of them (Vayikra Rabbah 4:6).

We have a responsibility to life, to defend it everywhere, not only against our own sins, but also against those of others. We are now all passengers, together, in this same fragile and glorious world. Let us safeguard our rowboat—and let us row together.

Rabbi Arthur Hertzberg

FROM THE MOSLEM DECLARATION:

Unity, trusteeship and accountability, that is *tawheed, khalifa* and *akhrahd*, the three central concepts of Islam, are also the pillars of the environmental ethics of Islam. They constitute the basic values taught by the Qur'an. It is these values which led Muhamad, the Prophet of Islam, to say: "Whoever plants a tree and diligently looks after it until it matures and bears fruit is rewarded," and "If a Moslem plants a tree or sows a field and men and beasts and birds eat from it, all of it is charity on his part," and again, "The world is green and beautiful and God has appointed you his stewards over it." Environmental consciousness is born when such values are adopted and become an intrinsic part of our mental and physical make-up.

Moslems need to return to this nexus of values, this way of understanding themselves and their environment. The notions of unity, trusteeship and accountability should not be reduced to matters of personal piety; they must guide all aspects of their life and work...

Dr. Abdullah Omar Nassef

The Shakertown Pledge

Recognizing that the Earth and the fullness thereof is a gift from our gracious God, and that we are called to cherish, nurture, and provide loving stewardship for the Earth's resources, and recognizing that life itself is a gift, and a call to responsibility, joy and celebration, I make the following declarations:

1. I declare myself a world citizen.

2. I commit myself to lead an ecologically sound life.

3. I commit myself to lead a life of creative simplicity and to share my personal wealth with the world's poor.

4. I commit myself to join with others in reshaping institutions in order to bring about a more just global society in which all people have full access to the needed resources for their physical, emotional, intellectual, and spiritual growth.

5. I commit myself to occupational accountability, and in so doing, I will seek to avoid the creation of products which cause harm to others.

6. I affirm the gift of my body, and commit myself to its proper nourishment and physical well being.

7. I commit myself to examine continually my relations with others, and to attempt to relate honestly, morally, and lovingly to those around me.

8. I commit myself to personal renewal through prayer, meditation and study.

9. I commit myself to responsible participation in a community of faith.

Shakertown Community,
Philadelphia, Pennsylvania

Other Ways to Pray

These celebrations are outside of times of formal worship. The examples are divided into two parts: meditations and retreats.

Guided Meditations

SEEING GOD IN ALL THINGS

The following celebration has as strong personal prayer aspect and includes silent meditation time either by a group or individual. It begins in a group through a shared prayer in which all place themselves in God's presence. The group then separates and the meditation is done individually. As individuals, they find the place, posture and circumstances most convenient and comfortable to them. Eyes can be closed if desired. At the conclusion, the participants reassemble and again pray together, either with a vocal closing prayer, a hymn, or other fitting closure. An alternative format would have the entire time taken in slow readings and reflection, with shorter periods of silence as seem proper. The following is a meditation:

> Of all the symbols in your spiritual life, light is the most crucial. Without light, you would stumble in the darkness and not know the way. Without light, the photosynthetic process would not occur and plants would not grow. *(Pause)*
>
> Light allows you to see and know. You may often say, "I see," and mean understand. Your eyes are your instruments. Without eyes ordinary functions of life become quite difficult. Your eyes distinguish objects for you, help you determine distance, and give you the chance to see beauty and color. And your eyes require light. *(Pause)*
>
> Seeing God in all things is the beginning point. Reflect upon the goodness of creation. Take yourself back to that grand beginning instant 20 billion years ago. In a fraction of a second the great actions that shape our universe occurred. Even then you were present in the Mind of God. And God saw all of creation as good. *(Pause)*

Gaze to the stars and see the pinpoints of light which started on a journey millions and billions of years ago. Become breathless in wonder in beholding Earth's candles giving the Creator praise. Look under your feet and penetrate into the world of small insect and bacterial life, another vast universe. Here you stand between these two vast worlds. *(Pause)*

Seeing God in all things can make you aware of the importance of your physical eyes and also your eyes of faith. Thank God for the eyes of faith. God, the Creator, reveals divine attributes and qualities in the various plants and animals and Earth formulations around us. Rediscover these divine attributes in creation here and now. Resolve to see all Creation as God's Manifestation of Self to you.

PEACE WITH NATURE (Meditations for the Young)

We were not here when the Lord, wise and powerful,
made the universe out of nothing,
and when God made the Planet Earth.

God made the light, the air, the water, and the land,
with minerals and other hidden treasures.
God made the plants and the animals,
all sorts of living things,
all beautiful and interesting.

When there were no people on this earth
the air, the waters, the soil, were clean and healthy
for all the living beings to share.

All the living creatures and all forms of matter
the Lord made gave glory to the Lord
just by being what they were.

God created human beings in the divine image
to be witnesses of divine love.
God wanted us to be happy and to be free:
to use our free will participating in God's Divine Plan,
lovingly caring for all Creation,
common inheritance for all people of all times.
People came into a world of clean air, waters and soil,
full of different forms of plants and animals
to live among them and with them as a very special form of life.

The Lord gave us intelligence and skills
and ways to know right from wrong.
God gave us the task to care for the Earth
and to make it more useful and helpful to each person and to
society through science and technology,
so that we could all live in peace.

When there were no people on this Earth
there were no wars.
Plants and animals followed the laws of the Lord
serving one another in their life and in their death.

People made wars on nature
and they made wars among themselves.
And now we have an environmental crisis:

But people did not recognize that they were placed on the
Earth to care for the heritage of God who made
everything out of nothing.

Air and water pollution, massive deforestation,
fertile lands turned into deserts,
important plants and animals
in danger of disappearing forever,
disasters caused by human errors.
The living beings that occupied the Earth
before we came did not cause all this damage.

Many of them did not survive
and others are in serious danger of extinction
because people are constantly at war with nature
and what we need is to be at peace.

War with nature is a war we cannot survive.
We must do everything we can to stop it.
Science and technology must not destroy nature.
Knowledge must be used to save nature
and to restore it.

Forests, soils, waters, the fauna, all are necessary.
They must be used wisely to permit their renewal.
We cannot use them without considering the future.
Future generations, after us, will depend on what is left.

We have duties towards God and towards our neighbors.
We also have serious obligations towards nature.
Our life depends on it and we owe nature a great debt.

Let us contemplate nature undisturbed and at peace:
the skies and the mountains,
the sea and the rivers,
the fields and the forests.
Let us look at the beauty
of the little corner where we live.

There are still many beautiful places on Earth
that we can protect and make more beautiful.
There are still many plants and animals
we can save from extinction.
They may help us to survive.

All the things the Lord made are wonderful:
all are good, all are beautiful.
God gave us all these things to care for, to share,
to protect from destruction
and to restore in abundance and beauty.

The Lord gave us the intelligence
and the capacity to love.
God also gave us the grace to do this with joy:
to restore the Earth so that we may live in peace with Mother Nature.

From the Children's Spring Festival "Peace with Nature" (18-21 March, 1988) by Sophie Jakowska, Ph.D., Member IUCN Commission on Education and Ethics Working Group.

Address: Arz. Merino 154, Santo Domingo, Dominican Republic. Illustrated with slides by: the author, Planning for Survival IUCN-ICCE, Plan Sierra, INDESUR, CARITAS, Dominicana, & Revista Maryknoll. Total 68 slides.

Retreat Ideas

There are many different kinds of retreats that can help people look deeply within and outside of themselves. Retreats can take place in a church or in a natural setting. Some retreats emphasize a feeling of kinship with Earth, others get at a secondary question that engages the deeper consciousness: What must I do to save the Earth?

For a copy of *Sacred Environmental Songs* and a book of reflections entitled *Appalachia, A Meditation*, write: Appalachia— Science in the Public Interest, Box 298, Livingston, KY 40445.

HIKING RETREATS

Guided or unstructured tours singly or in groups within natural settings may be ideal times for prayerful reflection. A week retreat is held during warmer weather with about eight persons. The participants carry their own equipment or have the equipment dropped off at the end of the day by a non- participating party. The group has a director who conducts a morning and evening prayer and a worship service that includes the sharing of trail reflections as part of an organized campfire chat.

A hint from experience: Cover only as much ground in a day as the weakest member is capable of doing, otherwise hiking becomes the major concern. Understand that for some, long walks such as the entire Pacific Crest Trail are appropriate, while for others, two miles is sufficient. Form your groups accordingly.

A NATURE EXPERIENCE

This kind of retreat may be an annual routine or a first venture. It includes a "nature immersion" that allows the creatures of the Earth to speak to the participants. This immersion will create an "Earth-consciousness" tone. The setting may be quite primitive: some people may not even take a tent. Fasting from a portion of food, alcohol, and tobacco is often recommended. Harmonizing with nature may require isolation from other people and times of silence. Often a coordinator is part of the experience.

WEEKEND AWARENESS EXERCISES

This retreat experience focuses on Creation- or Resurrection- centered spirituality. It includes sharing experiences, nature walks, and drawing exercises. A variety of environmental aids such as art, posters, sample prayers or decorations are helpful for these programs. Often this kind of retreat is geared to persons working in the area of social justice who have become aware of environmental concerns. (See also "The Earth Community Center" on page 48.)

SACRED SWEATS

The American Indian practice of a "sweat lodge" is another way of getting in tune with Mother Earth. In a small hut, tent, or temporary building, very hot rocks are placed at the center on the dirt floor. Participants sit around in silence or tell stories, chant or sing songs. The purification is a way of meeting the Earth and can be preceded or followed by liturgical celebration of a variety of models.

PREACHED RETREAT

For this retreat, individuals set aside a moderate period of time (e.g., six days). During that time, someone gives a series of talks alternating with times of silence and reflection. One idea is to move through the liturgical cycle of suggested readings. A special part of creation (e.g., trees, flowers, clouds) could be the focus of each session and, by means of narration, the way these elements fit into salvation history could be shared.

RETREAT DAYS

These days may form a series within the church year (e.g., monthly) or part of a particular church season (e.g., Lent). They can have featured speakers and others who assist in songs, dramatic readings, confession rites, or sacred dance. These days can include reflection and discussion on several topics: personal involvement in environmental issues, the requirements for a balanced activism, the struggles of the present, and the dreams for the future. These days do not require the expense of retreat buildings and staff.

NatureQuest

NatureQuest offers eight days in a wilderness area alone but under the guidance of trained leaders. There is no formal program but a guide provides basic training in outdoor living and ways to assist in deepening your experience of aloneness. Each person lives as close to Nature as possible, without artificial light, reading material or unnecessary and distracting material comforts. Locations throughout the country. For information, send a $3.00 donation to: John P. Milton, NatureQuest/Drawer CU, Bisbee, AZ 85603 or call (602)432-7353.

An Environmental Retreat

The Environmental Retreat consists of twelve topics taking a group through the liturgical seasons of the Earth. It is a journey of faith and renewal that allows participants to see that the Earth suffers. It also helps the retreatants appreciate that through their suffering they share in the work of Christ — the redemption of the world.

The following is a list of topics used for the Environmental Retreat. Other retreats are also available.

1. Seeing God in All Things
2. Winter of our Discontent (Sufferings)
3. The Role of Being a Suffering Servant
4. Resurrection
5. We, too, are Resurrected People
6. Ascension: Being Stewards of the Earth
7. Pentecost: Being ourselves in a Hostile World
8. We are the Church
9. Transfiguration
10. Entering into the Total Liturgy
11. A Reflection on our End of Journey
12. Thanksgiving for God's Gifts

Information on group retreats is available from Appalachia–Science in the Public Interest, Box 298, Livingston, KY 40445.

The Earth Community Center

Justice for the People/Justice for the Earth:
Two Sides of the Same Coin

Jane Blewett, Director of The Earth Community Center, will offer workshops and retreats at your location. She weaves together some of the essential themes of Catholic social teaching and today's urgent ecological concerns.

Linking faith and justice	Political participation
Dignity of the human person	Promotion of peace
Option for the poor	Economic justice
Global solidarity	Stewardship

How can we relate these traditional themes of Catholic social teaching to the current, widespread damage to our air, water, soil, forests — our life-support systems?

How can we expand the social agenda of the Church to include not only the human community but the total earth community?

Is it enough to make an option for the poor unless we also make an option for preserving the earth that gives life to the poor and to all others? What do eco-feminists say about the destruction of the planet?

Workshops and retreats feature:

Presentations	*Video Tapes*
Discussion time	*Music*
Personal prayer	*Slide/tape shows*
Reflection	*Suggested bibliography*

The Earth Community Center
15726 Ashland Drive
Laurel, MD 20707 (301) 498-2553

DIRECTED RETREATS — For this kind of retreat, individuals or a small group move through a series of exercises leading to a particular resolution. For example, participants may want to find what area of environmental work is most suited to them. With the aid of a director, they enter fully into the discerning process with the person during this retreat and beyond. This type of retreat can be made anywhere but natural settings are usually better for an environmental retreat. God speaks to us in the sights, sounds and smells of the Earth.

ANNOTATED RETREATS

This retreat does not differ significantly from the preceding, except that it extends over time. It is actually a period of time set aside in the working week. A person follows a prayer/direction routine with a special emphasis on environmental awareness or Earth concerns. Retreat or spiritual directors can help facilitate the annotated retreat over a given period with meetings either as small groups or on an individual basis. It is a retreat for those who cannot find time for specific directed retreat periods during the year.

RETREATS IN PREPARATION FOR DIRECT ACTION

Radical forms of environmental action should be preceded by prayerful reflection on the particular non-violent action. This retreat requires soul-searching and decision-making. Radical action may involve being arrested, misunderstood, or losing support or funding. The retreat focus is not on the decision to enter the environmental movement, but on the effectiveness of an individual's personal witness. Here the non-violent workshop leaders from the peace movement have a fertile field since little has been done in this area. (In fact, the church may be called to bless radical environmental action.)

HERMITAGE EXPERIENCE

Sometimes an individual desires to spend a period of time alone and living close to nature. This can be done in a "hermitage." A hermitage setting may range from a little piece of suburbia in the wilderness to rather primitive quarters with few amenities in the city. For this retreat, the Earth itself is the teacher. Time is needed to sort out living problems, to pray, and to reflect on the actions to take to renew the Earth.

STRATEGY SESSIONS FOR PARISH ACTIVITIES

For this experience, a planning group meets periodically (e.g., twice a month) to discuss ways to mobilize resources for a particular battle such as a hazardous waste landfill. The meeting includes prayer, a communal sharing of Scripture (similar to a base-community experience), and sessions on organizing, church involvement, and strategizing ways to bring about change. Through this retreat, the church becomes the focal point for change and assists the environmental groups at the local level. The church also enables the participants to find their own niche. (See Chapter IV for more information.)

A Weekend Retreat

Inner World/Outer World,
Life through the Eyes of a Poet and a Scientist

The focus of this weekend retreat is exploration of the interconnectedness of the objective world of nature. This is achieved through poetry of the inner world of feelings and personal reactions to experience and through observation and discussion. It combines poetry readings, a small group process in which feelings are put into poetry, and a nature walk (emphasizing the interrelatedness of the organic and inorganic world as seen in a forest or some other setting).

The entire group listens to the poems that resulted from the small group writing session. One or two slide shows reinforce the observations of natural systems. The final worship service, includes the topics' interrelatedness and the role humans play as stewards of Earth. In terms of organization, the sponsoring church provides adults who insure that children in the group have recreational activities, arrange for meals, sleeping, etc. For information about this retreat: John and Grace Freeman, Route 4, Box 232, Brevard, NC 28712, (704) 883-3443.

III. A Critical Look at Lifestyles

Being concerned about the Earth demands a critical look at lifestyles: both of church leaders and of the worshipping community. The Western lifestyle has now spread through much of Europe, the Pacific Rim, and parts of the Middle East. It is a way of living marked by affluence and consumerism and it devastates the Earth and its people in direct and indirect ways:

- **through depletion and waste of natural resources.** The average American uses about one thousand times the non-renewable resources as does an average Nepalese.

- **through environmental pollution stemming from processing and distribution of consumer goods.** The average First World person has over ten times the impact on the world's environmental quality as does an average Third World person.

- **by means of military activity used to defend and preserve extravagant lifestyles.** The Gulf Conflict was in part precipitated by a desire to retain a steady flow of cheaper oil mainly for First World countries.

- **through the insensitivity of affluent people to the world's poor and destitute.** Affluent living is filled with many distractions that distance "the haves" from the problems of "the have nots."

- **through the loss of faith in the co-creational abilities of the believer.** The consumer distractions not only squelch creative abilities, they deny people the very experiences that nurture faith.

Lifestyle Change

A Necessity for the Affluent

You as a church leader stand at a crossroad in your service to the worshipping community. The consumer problem and lack of a balanced evangelistic response affects your pastoral role. An unchallenged affluence can undermine the faith that you are trying to nourish. To challenge it, however, may lead to divisions and friction among the members. To fail to address it in the your own life will open you to charges of hypocrisy.

The enticements are strong to overlook the lifestyle question. But church leaders who recognize that our Earth is in trouble must begin to change their own lives and to lead their worshipping communities beyond a mere conscious awakening to the next two concrete steps: assessing their personal use of resources and instituting lifestyle changes.

Getting "unhooked" from a consumer culture is not a simple rational exercise — not in a world where billions of advertising dollars are spent reinforcing the consumer addiction. Miracles are possible, and converting some "super-consumers" may demand a miracle.

While this chapter is appropriate for all members of the consumer society, more of the material deals specifically with the lifestyles of middle and upper classes. These groups use a far greater amount of resources than the great majority of the world's other consumers. Much of their use, whether for speed boats or jet fuel, goes for pure pleasure and self-advancement; they are luxuries that can be mistakenly redefined as needs. The same can be said about wasteful misuse of non-renewable resources or the common practice of keeping homes a little warmer in the winter than is considered comfortable in summer. In such a climate, you must strive to awaken your congregation to the insensitivity and selfishness of such practices.

The following are procedures to consider before initiating lifestyle reflection and change.

- **A resolution by the church leadership to simplify their personal lifestyles and the lifestyle of the community.** This may involve recruiting the services of a respected person who is both sensitive and articulate about simple living. The person can serve as an admonitor or spiritual director.

- **A direct and radical confrontation during dialogues, sermons, and other communications about the serious threat to faith created by affluence and wasteful lifestyles.** Youth are generally more willing to question their own lifestyles than are adults. It may be best to confront some members of a family, i.e. youth, and let them take the need to simplify to the rest, i.e. adults.

- **A group process that engages others in an audit of their own individual and community lifestyle.** Here facilitators prove valuable. They can find out which questions to ask, which require more social awareness, and which are to be postponed because they are threatening. Questions which reveal just precisely how much individuals misuse or waste certain resources can be difficult to work through.

- **Introduction of Alcoholic Anonymous methods to the lifestyle simplification process.** It is possible that people who have been helped by the A.A. methods are available to the church community. They, as well as others, may be challenged to apply A.A. methods to the problem of over-consumption. The A.A. person sees the need for help from outside power and recognizes that often one is unable to make drastic personal changes alone.

- **Group discussion of lifestyle simplification options:** aggressive governmental redistribution programs through political means; revolutionary or conflict-oriented events such as Kuwait in the aftermath of the Gulf War; the choice of doing with less; the forced acceptance of doing with less because of unforeseen events such as financial collapse.

- **Group meditation.** See "A Reflection on Lifestyles" on page 66.

A Different Way of Living

Modifications are needed to simply live in the present world. While there is no absolute standard of model behavior, there are ideals worth striving for. All other things being equal, people with simple lifestyles are those who:

There are no hard and fast rules of do's and don'ts. Biking uses no gas but takes time and is difficult in inclement weather. It is a healthier and safer mode of travel, except when commuting with auto traffic. For a person who must travel to areas where public transportation is usually unavailable or infrequent, justification for a car exists.

- *Eat moderately and lower on the food chain.*

  Eating less is a health tip in a society blessed with plentiful food and in a society where obesity is a problem. Limiting food intake is an obvious start. Eating healthy food lower on the food chain means eating foods that require fewer resources to produce (unlike overly processed "junk food"). "Eating lower" means equal or better nutrition; only the choice and amount of each type of food will change. Eating lower means less meat (especially red meat) and more whole grains. It means buying locally produced fruits and vegetables. In terms of a final food product, a pound of beef from grain-fed cattle uses over 20 times the resources that a pound of whole grains and fresh produce does.

- *Live in simple, safe, energy-efficient dwellings of native materials.*

  The important points of harmonious home environment are: economy, household safety, and low environmental impact. Trade-offs are obviously part of the total picture but without considering all three, the home becomes the most non-regulated part of the total environment.

  Economy means adequate space for the number of persons present. Safety means using safe chemicals in the indoor environment. For energy efficiency, the interior space should contract (e.g., close off porches and patios) in winter and expand in warmer weather. It also includes both conservation practices (such as adequate insulation) and use of renewable energy sources such as sun and wind for space heating, water heating, and electric generation. For much of the world's housing native materials are easier to obtain and transport. Bringing bulky materials from great distances is outside the price range of many people and such practices expend limited non-renewable resources as well.

- *Wear clothing for durability and comfort.*

  Clothing keeps wearers warm or cool and protects them from the elements. While fashion has some value, it cannot be all consuming. Neither can fashion mean the wardrobe is emptied at every change of season. A simple lifestyle includes obtaining clothes through trading, rummage sale, flea market, and other outlets. The same can be said of other personal items such as home furnishings, decorations, curtains, rugs and bed coverings.

  Some consider it important to give away unused or unfashionable clothing at an annual clothing drive, in order to purchase new items. Each person should consider clothing needs as primary and fashion as secondary. Teach others in the family to do the same.

- *Use land for food growing, not simply for decorative purposes.*

  Over half of North American households grow plants. Much is of a decorative nature. Edible plants can be both productive and beautiful. Edible plant gardening can save money, yield nutritious fresh produce, afford the gardener needed exercise, fresh air, and an opportunity to touch the Earth. It saves transportation costs and refrigeration energy. Learning how to garden, what to grow on the land available, when to plant, and how to preserve surplus produce, is all part of becoming self-sufficient and thus living more simply. Home-grown produce out of season saves the precious energy required to bring the same food (perhaps exposed to more chemicals) from distant places.

- *Re-use waste within the boundaries of the immediate environment.*

  Through creativity people can produce less garbage: buying in bulk to avoid massive packaging and preserving home- grown produce. Compost toilets (where allowed) can compost human, kitchen, and yard wastes. The compost product can then be returned to the growing areas. What cannot be re-used on the premises (aluminum, paper, glass, rubber, waste oil, other metals) can usually be recycled. Excess rain water can be stored in cisterns for use on the grounds.

- *Reduce private automobile use.*

  The American culture promotes the two-plus car family. Virtually every activity requires a motor vehicle. Some errands can be done by walking, biking, or with a small vehicle. If an automobile is needed, the people who live simply try to car pool and they drive with energy efficiency in mind. They keep their car well-maintained and take vacations close to home (or combine vacations with a business trips).

- *Undertake recreational activities that utilize few non-renewable resources.*

  Much of the home entertainment of the consumer culture is from televisions, stereos, VCR's, Nintendos, and similar devices. By returning to participative games and music, resources and electricity are saved. The savings occur also through the conservative use of electrical household appliances and lighting (or by refraining from unnecessary purchase in the first place).

  Modern outdoor recreation includes vacation travel, trips for entertainment, and participatory or spectator sporting events. The amount of energy used in each differs enormously, from nearly none (jogging near home) to a great deal (skiing at a distant resort).

- *Fulfill health, educational, and personal needs near or at home.*

  Travel to distant institutions for medical, educational and other services often can be omitted if the service is performed at or near the home or in a local setting (community health clinic, community college, or home nursing service, etc.). Granted there are good reasons for using distant institutions, however, home or decentralized facilities generally use fewer resources.

- *Help create cooperative community activities.*

  A community is rebuilt when groups who desire to be self-sustaining unite to do things for each other, such as sponsoring of food, work and service cooperatives, creating land trusts, community gardens, and community marketing ventures. Cooperative activities keep capital within the local area. These activities check the drain of resources to outside centers of power. Social investment that assists local groups is vitally important. Earth-renewing community activities include such things as creating wilderness areas for wildlife, tree-savers associations, and bird sanctuaries.

How-to books, which explain ways to conserve and to achieve a simple lifestyle (with respective pitfalls of each method), are easily found. By becoming aware of resource conservation, conscientious consumers soon see which exercises save and also fulfill the community. The challenge remains: how can we improve socially with the least expenditure of world resources? The person who lives simply has this challenge always in mind.

How Much Do You Use?

The cartoon character "Cathy" was asked by the grocery clerk whether she wanted paper or plastic. She answered, "Paper which decomposes in a week or plastic which decomposes in 400 years? Of COURSE I want paper." To which the clerk responded, "Fine. I will pack your plastic sandwich bags, freezer bags, trash bags, household cleaner bottles, detergent bottles, and soft drink bottles in a nice biodegradable paper bag." She added in the last frame of a blushing Cathy that it takes fifteen seconds for self-righteousness to decompose.

In this section you can assess your own lifestyle. You can also use these questions as an audit for worshipers. Checking off a list of "do's and don'ts" is never sufficient, but like the church audit, assessments are a starting point — and an important one.

The focus of this assessment is the amount of non-renewable energy each person uses for nourishment, housing, clothing and personal items, yard care and waste disposal, travel, recreation, health and personal services, and community activities. The resulting answers put a perspective on the kind of changes you will need to encourage. Remember that many questions from the church audit in Chapter I are applicable to individuals or can easily be adapted.

Food

A major consideration in food auditing is the number of people that comprise the family or group living unit. A basic principle is that the fewer mouths to feed per meal, the more energy required per person to do so.

☐ How many eat on a regular basis at your residence?

☐ Do you have at least one cold meal a day (particularly in warmer weather)? Do you cook dishes in large enough quantities for use during a number of consecutive days so that only warming is necessary?

Eating a cold meal is a major energy savings.

☐ How do you cook? In a very environmentally benign solar food cooker occasionally? On an outdoor grill? On a wood burning stove? A gas stove? An electric stove or hot plate? A microwave oven?

The microwave uses far less energy per operation performed than does the gas or electric range.

☐ Do you use little pre-packed convenience food (that requires energy to produce)? Do you cook little or no red meat? Do you save wilted and left over vegetables for soup or stew?

Almost half of all produce shipped to market is never eaten.

Housing

The type, size, and condition of your home can determine major resource expenditures. Because more people now move away from nuclear family units, in the last two decades the number of dwelling establishments — and the amount of energy it takes to maintain them — has grown twice as fast as the population.

☐ In what type of building do you reside? How many others live in this building? Has this number increased or decreased in the past decade? Is the space adequate for the residents' needs?

☐ Is the building adequately protected for both winter and summer with proper insulation and shading? Does it need further weather protection? How is it heated in winter? Cooled in summer?

☐ Have you considered using solar energy? Does the building have areas that can be opened for more space in summer such as porches and patios?

☐ What is your residential water use? How do you heat your hot water? Have you considered solar water heating? Are there additional ways of conserving water? Do you re-use your gray water? Can you have a compost toilet?

- [ ] Are you careful to preserve the indoor air quality of your home? Do you prohibit smoking? Do you use chemical cleaners? Are there chemical hobby supplies? Is the home properly vented?

- [ ] Have you created quiet space for those in the family who seem to need it most? (See "Home as Sacred Space," page 77.)

- [ ] How many electric appliances do you have? Are there more than 40? Are some unneeded (e.g., electric pencil sharpeners)?

- [ ] Do you have a self-defrosting refrigerator, a big energy user? Can you live without some appliances?

- [ ] Are you watchful about the amount of artificial light you use? Do you buy energy-saving bulbs?

Clothing and Personal Items

Clothing audits are most revealing. You may not imagine how much you spend on clothing unless you have a very close budget. Almost everyone is surprised by the amount spent on personal items, ranging from mementos and trinkets to cosmetics and personal care products.

- [ ] How much do you spend on clothing/personal items each year? Are you willing to keep track?

- [ ] Do you ever use hand-me-downs? Clothing purchased at garage sales or flea markets? Do you give away clothing as charity in order to vacate the wardrobe for the next fashion change? Are you hard on clothes or take care of them?

- [ ] Are you knowledgeable about which fabrics are natural and which are synthetic? Are clothing and shoes purchased for fashion or for durability and ease of maintenance, cleaning, and repair? Do you acquire for quality?

- [ ] How much of your money, maintenance time, and space is consumed by these items each year? Are these items an investment, a necessity, a luxury, a status symbol?

- [ ] Are you someone who avidly collects mementos and trinkets? Do their care and the space they consume bother others?

Yard Care

Your home's exterior space is a major resource consumer. You may or may not have control over this space. If you do have a say, it is helpful to ask some some questions about this part of your lifestyle.

A Critical Look at Lifestyles

- [ ] Do you cultivate part or all of the lawn area? Do you allow some of the lawn area to return to wild flowers and wilderness so that you need not mow or upkeep the lawn?

- [ ] What portion of your food needs are home-grown? Are you resolved to increase this in the coming year?

- [ ] Do you collect your own rain water and use it to water plants? Do you have a solar greenhouse, seasonal extenders, or cold frames?

- [ ] Do you use commercial chemical herbicides and pesticides in any manner?

- [ ] Are you mindful of the need to reduce, sort, compost, and recycle your wastes? How much do you generate each week for a land fill?

Travel

Your travel choices affect resource expenditures and are usually based on such non-environmental factors as time saving, comfort, privacy, convenience, and fashion.

- [ ] Do you have a car? Would you classify it as energy-efficient or a gas guzzler? How much do you drive it each year? How much is pleasure? What part is work and services for others?

- [ ] Do you drive your car at reasonable speeds? Do you maintain it well?

- [ ] Do you car pool or share the vehicle with others? How many use the vehicle?

- [ ] Are you the type who plans the travel route and arranges what errands can be accomplished in the planned trip?

- [ ] Do you ever walk, bike, or use a smaller car for local driving?

- [ ] Are your residence and work place accessible to public transportation? How often do you use it?

- [ ] Has public transportation use increased or declined in your life? Why so?

Recreation

The amount of energy used for recreation varies significantly from person to person. No one can prove that greater enjoyment or health comes through greater resource use. Since there are good, environmentally compatible recreational activities, an audit may allow you the opportunity to change.

- ☐ Do you prefer participative rather than spectator sports and activities? If not, do you spend much time watching televised events? Could this be reduced by having other forms of entertainment?

- ☐ Are you an electronic-oriented person (stereo, VCR, etc.)? How much money and time have you spent in the past five years on electronic devices? Do you use these at least once a week? Has this curtailed your personal creative activities?

- ☐ What activities do you do outside? Are you a hiker, walker, biker, jogger, bird-watcher, or do you participate in any other low impact activities? Do you reach your outdoor area of exercise by foot or car?

- ☐ Are you willing to vacation closer to home or at home with some diversified activity? Have you considered an alternative vacation such as offering service to poor people?

- ☐ Do you respect those you visit when on a tour, for example, photographing only those willing?

Social Services

It is quite difficult to make an audit of social, health, and educational services because opportunities vary considerably depending on your geographic location and your economic status. Not everyone is free to make choices, therefore this section is limited to a few specific questions.

- ☐ Do you make use of the social services in your community (e.g. family counseling centers, crisis centers, hotlines, mental health services, etc.)?

- ☐ Do you belong to a cooperative medical plan? Do you have regular physical, eye and dental checkups?

- ☐ Do you assist in alternative educational endeavors in any fashion? Do you make plans for your own continuing education?

Community Activities

If a healthy community is at the heart of a renewed Earth, church leaders must be foremost in community public interest activities. The inconsistent record of community participation by church leaders may have justifying reasons. You know leaders in the church who find such time. Let them become models for others. Support them at gatherings when they are present or when their name is mentioned.

- ☐ Do you regard yourself as a "community person"? Are you registered and do you vote? Have you served on the jury? Are you present and participate at community meetings and festivals? Do you foster participation in environmental monitoring or advocacy?

- ☐ Are you a member of a local land trust or cooperative—food, growers, marketing, etc.?

- ☐ Do you volunteer time to work in some agency or non-profit group in your community? Are you involved in advocacy work for the poor in some fashion?

- ☐ Do you give time to environmental groups? Do you encourage others to strengthen and beautify the neighborhood? Is land in your area being set aside for parks, wilderness areas, and bird sanctuaries?

What Happens to Your Time?

The audit may make you uneasy. It is obvious that living simply takes time — often more time than convenience-driven lifestyles. But most people feel there is never enough time.

Where does your time go? The key is budgeting your time, your most precious possession. People who are good planners are able to use time efficiently and be healthy and in harmony with the Earth. The following personal time budget may be the beginning of a more critical look at where your time goes.

Note: An alternative approach to "lifestyle assessment" is found in the appended analysis of simple lifestyle practices at the end of the book. It shows the wide range of benefits that result from these practices. You may wish to study these pages very carefully.

Personal Time Budget

Record the time you spend on different activities during an average week. One way is to record your time is to record it according to the nearest whole hours (168 per week). Omit any extraordinary weeks in your calculations (vacations, holding family events or holidays). Some time can count double (e.g., reading while traveling).

Basic Personal Activities:	Average Simple Living:	Your Amount:
Sleeping and resting (7/day)	49	
Eating at (2/day)	14	
Traveling:		
To and from work place	5	
To recreation	1	
To market	2	
To worship, etc.	2	
Personal cleaning (1/day)	7	
Clothes, dishes, room cleaning operations	2	
Work (direct)	40	
Preparation & indirect	10	
Gardening & outdoors (1/day)	7	
Television	5	
Reading (all times)	7	
Worshiping, praying & meditation	7	
Sports, exercise & recreation	7	
Planning	3	
TOTAL	168	168

The Affluence Cycle

Consumers use and demand more goods. Corporations respond to increase their profits; create more goods. Government is complacent. Local resources are strained.

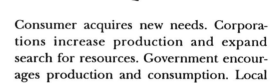

Consumer acquires new needs. Corporations increase production and expand search for resources. Government encourages production and consumption. Local resources become exhausted.

The poor and activists move to destabilize the state. Corporations call for military dictatorships and more arms. Governments become less democratic. Further erosion of liberties and resource depletion causes collapse of local structures.

Poor in other lands suffer from lack of basics. Corporations pressure resource-producing lands to imitate the First World and consume more goods. Third World governments become securers of resources.

Peer pressure increased consumer demand. Corporations extend search to other lands. Government becomes protector of corporate market/resource base. Communities are damaged.

Community structures break down in Third World. Differences in abilities of various consumers to gain materials become more pronounced. Corporations pressure governments to secure markets/resources overseas. Governments set up more strike forces, military bases; enlargement of arsenals. Communities lose control over production and distribution.

Group Discussion

Personal Lifestyle

- What will you do with the information gathered in this assessment?

- Are you overusing some resources? Which ones? Do you know why?

- How much responsibility do you assume to conserve resources for future generations?

- Does your lifestyle reflect the belief that the world will last for many more generations?

- How would you express the relationship between non-renewable energy use and your respect for the Earth?

- How can others benefit from the savings gained because of your resource conservation? Through direct sharing of saved resources? Are there any other ways?

- How is your environmental awareness expressed in conservation practices?

- How do institutions in your neighborhood contribute to the total community environment?

- Is your personal means of spending and acquiring money environmentally compatible? In what ways?

- Are you rich by world standards?

- Are you willing both to simplify your lifestyle and to assist others in obtaining the basics of life?

- What motivates your choice of livelihood, your means of saving, your methods of spending?

The earth provides enough to satisfy every person's need but not for every person's greed.
—Mahatma Gandhi

A Critical Look at Lifestyles · 65

A Reflection on Lifestyles

Use the following with your membership. Small group settings are best. Read these sections giving ample time for self-evaluation.

Quality of Life

Environmental problems are not only "out there." The Earth suffers because of what *we* do (or fail to do), what we give and take from the Earth, what we teach our children, how we live, work, and recreate. The Earth's healing demands our healing and that healing rests on our willingness to embrace the spiritual values that are the heart of our religious traditions. Through self-examination we see how to respect the Earth. In knowing our relationship to the Earth, we find her to be a good teacher. She leads us to deeper self-respect and respect for our Creator.

We in the West pride ourselves in our ability to live as we choose. Freedom does not apply solely to speech, the press, and religion, but to the fundamental right to fashion our lives and environment to meet our goals and aspirations. How free are we?

Lifestyle may be defined as outward manifestations of inner values. It also comes from the way our peers expect us to live. Our lifestyles are subject to some or much coercion, depending on our conformity to external pressures. Where are we on the freedom scale? Generations of families have worked to afford better homes, more comforts, and education for their children. In the name of progress, they have made compromises and trade-offs along the way. Has this seeking for a better life reached a dead end? What legacy will we pass on to our children today — the ability to destroy this planet with weapons, irreparable damage to our waters and land, and only a handful of memories of good times spent together? What values do our lifestyles express to future generations?

Charity— or Justice?

What material things are truly essential? Proper nourishment, clean water, clothing, and housing, as well as access to land, health services and public transportation: these are the necessities of life. People on a food line in Ethiopia barely have enough resources and they must live for that day alone. Choice barely enters their lives. Each day 40,000 babies die from hunger or diseases. Two-thirds of the world live at or below the poverty line.

But one-tenth of the population, the majority coming from the West, live in a state of affluence. This small group uses half of the world's resources. In the United States alone, more than a billion pounds of pesticide are used each year. About 3.5 million trees are cut daily to supply paper and building needs. The United States creates a ton of hazardous waste per person every year, some of which is transported to rural areas, countries of poor people, or dumped in the ocean.

Quality of life erodes when resources are scarce and when people are unable to become self-sufficient. Quality of life also erodes through an overuse of resources. Over consumption depletes scarce material and increases pollution. It also detracts from matters of the heart and, like excessive and costly defense budgets, it increases, not decreases, the human insecurity. When the haves and the have nots exist side by side, when the faces of the destitute are bypassed or ignored each day, all humanity goes on trial. What is the verdict?

Creating an Environment

Balance is to be sought. Our task is not to delineate the ideal means to achieve equity in the world. It is to encourage worshipers to live more appropriate lifestyles. All lifestyles require material goods but how much is used and with what attitude can vary.

Consider the distinction between a consumer lifestyle and one that conserves resources. With the former, the material goods matter. The amount of possessions is often more than the individual needs or can use properly. This lifestyle obliges a person to earn more. It requires time to maintain and protect acquisitions. It diverts a person's attention to the maintenance, security, and replacement of material goods and can easily warp the sense of real need. Often it affects other creatures — human, animal, vegetative — negatively. Freedom and creativity are limited because the person fails to respect creatures and thus co-creativity with them.

A conservation ethic is one that values goods and their proper use. It is one that respects God's gifts and cares for these gifts. Security comes from the well being of the entire community (the distant needy as well as immediate neighbors) not from material possessions. Freedom and creativity are enhanced by conservation because opportunities often arise that enable people both to appreciate and to make better use of fewer goods.

Every creature affects the Earth. Everyone has a choice: to make small footprints that fade quickly without lasting marks, or to scar the Earth in ways that will be remembered for generations. A North American Indian proverb says, "We borrow land from future generations." How do we want to be remembered — as destroyers or re-builders of the Earth?

A Covenant Group for Lifestyle Assessment

PURPOSE

1. To consider lifestyle options in response to:
   (a) the hunger/ecology/justice crisis and,
   (b) the freedom and responsibility given by the gospel.

2. To agree as a group on specific changes that members will commit themselves to carry out for the duration of the group's existence.

3. To function as a support community enabling each member to see, accept, and enjoy the changes that lead to a more faithful, fulfilling style of life.

INITIAL COVENANT

1. To participate in each session.

2. To do some preparatory study and to share in the leadership of the group.

3. To express oneself honestly, and to listen carefully.

4. To be faithful in keeping promises to make specific changes.

AREAS OF LIFESTYLE CHANGE

1. Consuming
2. Conserving
3. Sharing
4. Playing
5. Advocating
6. Giving

The covenant group is a group of 7 to 14 adults. There are 12 sessions, each lasting approximately two and a half hours. Sessions include Bible study, discussion, activity, reports on readings, participation in the *IMPACT* and *Bread for the World* networks, and the consideration of and commitment to specific changes. Three of the sessions are open to members' entire families and include a shared meal and other activity-for-enjoyment. Adapted from *The Participant's Manual* of *A Covenant Group for Lifestyle Assessment*, by William E. Gibson, Church Education Services, The United Presbyterian Church, USA, 1978.

IV.
Healing the Earth: Options for Action

When a group of Salt Lake City sixth graders and their teacher took on a toxic waste site, no one thought that they would receive a standing ovation for their proposed environmental legislation at the Utah State Capitol a year and a half later. If every church or worship community took on one cause, just think of the difference it could make.

There are various actions local churches can initiate to help heal the Earth. Churches are often involved in soup kitchens, meals on wheels, shelters, thrift shops, and a host of other outreach programs. Environmental concerns can be added to this list of causes. This is the time to mobilize resources. Resources can be viewed broadly: people, money, expertise and skills.

It should be stated that many church communities find their capacity to give already stretched to the limits. Recognizing this, remember that the Earth's suffering is not separate from the economic and social problems that affect people everywhere. Compassion means "to co-suffer." In striving to relieve our shared poverty — a lack of clean air and water — we may find the seeds of renewal for all.

Mobilizing for Action

The following steps will affirm your community's responsibility to the wounded planet. Each is important in effecting change.

Select a Cause

Often the cause you choose will be dictated by circumstance. It may be as straightforward as sponsoring activities for World Food Day, assuming responsibility for a recycling center, or becoming an advocate for the clean-up of a local toxic waste site. Sometimes the choice is decided by the interests of your volunteers. Perhaps the plight of Third World countries whose resources are squandered on First World needs has motivated members. Another time you may have to choose one cause over many.

No matter what your choice, every cause demands a plan of action. It is important to assess your capabilities and resources before committing yourself and your community. Selecting a cause demands a balanced dose of realism and idealism. Weigh out the good and bad alternatives. Reflect on your motivation and how your talents — and those of your church members — can best be directed.

Build a Base of Support

You will need a core group of workers. This group may spontaneously form among your members or you may have to work through an existing social action group or go to your membership-at-large for volunteers. Do not re-invent the wheel. Two major forces in the environmental movement that can offer expertise and resources are national conservation organizations (see the *Organizational Resources*) and local grassroots organizations.

Once you have a goal and plan, contact your regional Council of Churches for their support in publicizing your cause. Consider joining forces with other churches in your area. Environmental problems are perfect opportunities for ecumenical efforts.

Encourage Volunteers

Believers are compassionate toward the less fortunate and seek to help them. At special times in life (often after college, during retirement, or at vacation time), churchgoers are willing to give a portion of their time to assist others. Ask your community to consider volunteering to save the Earth itself. There are not many jobs that pay to renew the Earth — though you can regard environmental protection as a form of national and global defense. Global military defense costs a trillion dollars a year. Tithing this for eco-defense would give a 100 billion dollar budget for starters. Church leaders can share the following steps with community volunteers.

Steps to take when volunteering:

- **Discern what is best for you.** Many choose the wrong cause and regret the loss of time and energy. Think about your own talents and then try to find an opening best suited for you. One excellent criterion is whether or not you will enjoy the work.

- **Talk over the position with a current or former volunteer.** You may either be reinforced in your decision or be more properly directed to another environmental concern.

- **Read and learn about the particular environment** you choose to volunteer in, the culture of the residents, and the means required for making effective change. In other words, don't go in cold. Before volunteering somewhere for an extended period, it is ideal to visit the area. Perhaps taking a vacation to the target place would be helpful.

- **Once you begin to work, be open to change.** Stand humbly before your Earth-teacher. Be willing to listen attentively and learn. Volunteers should neither solely give nor receive.

- **Pray for constant guidance** while preparing for and during your volunteer period.

- **Keep a journal** of your reactions and spiritual growth.

- **Be willing to communicate your experiences to others.** They may follow in your footsteps.

Spread the Word

Media coverage gives visibility to a cause. An editorial, a letter to an editor, an advertisement, a public service announcement, a debate, or a press conference are all methods that you can employ. However, do not limit yourself to these typical means of publicity. A Berkeley, California ecology group put "parking tickets" with a message about the automobile's contribution to air pollution on the windshields of parked cars. A group in Wisconsin handed out a mock million dollar bill to call attention to the federal tax money wasted on an underground antenna system constructed in their region by the U.S. Navy, known as Project Elf. On the back of the bill was room for a personal message to send to the appropriate people in Washington, D.C. These were effective attention-getters.

Educating Your Membership

The church is in a unique position to address the healing of the Earth, whether it be informing members about an environmental problem, recruiting them for a critical study, or directing them to meaningful action.

Start a Library

An environmental library may be a table in the back of the church or a book case in the Meeting Room. It is successful if it helps to focus members on specific areas of concern.

A good source is the North American Coalition on Religion and Ecology. It distributes educational and theological material. Write: NACRE, 5 Thomas Circle, Washington, D.C. 20005, (202) 462-2591.

Suggested titles for your library can be found in the *Organizational Resources* at the end of this book. There is a list of core books as well as key environmental organizations. Another resource is your local library. Ask them to subscribe to environmental magazines and order relevant filmstrips and videos. Take advantage of their collection to supplement your own.

Publicize your library in the church bulletin or start your own newsletter. Use public bulletin boards to display your material.

Sponsor a Discussion Group

There are not always simple answers when addressing the political, moral, and scientific components of environmental problems. But often what seems formidable to an individual is manageable in a group study.

The topic selection is important. National and international problems attract attention, but local problems have other advantages. With a local problem. your group can rely on first- hand experiences, the negative implications will probably effect them directly, and the options for initiating change are accessible. Involvement of individuals in local issues is the key to bringing about change in communities and society at large.

Once a topic is chosen, steer the group into formulating a specific goal. It should be one which will have some practical effect in saving the planet. Knowledge for its own sake is not the point here. Next, outline the plan of study. Seek help from local experts. Tap church social action committees or peace and justice centers.

Finally, share the fruit of your efforts. Post the group's findings on a bulletin board or in the church newsletter or publish a pamphlet which highlights pertinent information and suggestions for change. If your group is ambitious, they can formulate a position paper or host an "environmental awareness event" as suggested below. Anyone willing to attend a discussion group has made a serious time commitment. Utilize their energy well. For information see *Organizational Resources* and *Books of Interest* in the back of this book.

Host Environmental Awareness Events

Not everyone is willing to put in the time and effort required of discussion groups. How then can you impart information to other church members? How can you recruit other environment-conscious advocates or at least, create a more receptive constituent?

The opportunities from the pulpit, lecture podium, or classroom are never to be underestimated. Celebrations such as World Environment Day or Arbor Day are ideal opportunities to focus attention on the needs of the Earth. In Chapter II, ceremonies that are spiritual in focus were discussed, however, there is a place for strictly instructive events. Viable options include films, fairs, workshops, lectures, debates, forums, and exhibitions.

The key is to remember that effective learning best takes place when the real concerns of the audience are addressed. That usually means choosing a topic that effects the audience directly or one in which the audience can help to change. From these concerns you can branch out to national and international issues.

An educational event may be another useful time to tap local experts, organizations, schools, and universities. Your format will depend upon the interest level of your group, their lifestyles and perspective, as well as your topic. Whatever your topic and format you choose, be sure to have handouts available for people to take home.

The following are two of the most ideal days currently observed by large numbers of Americans for environmental awareness opportunities.

ARBOR DAY

Only one tree is replanted for every four that die or are removed each year. The Earth's tropical forests are destroyed at the rate of about 28 million acres a year (an area the size of Tennessee).

In 1872 J. Sterling Morton recognized the importance of trees for the Nebraska prairie land. He spread doctrines of forestation and conservation. Among his many accomplishments, Morton promoted an annual tree-planting day that gave prizes to the individuals and counties that planted the largest number of trees. His idea was so enthusiastically received that the first Arbor Day, April 10, 1872, over a million trees were planted.

Today Arbor Day is recognized across the country (although the specific date varies among states). Usually held in the spring during March and April, it is an opportunity to plant trees and to educate people about this major environmental issue.

Global Releaf

Global Releaf is a national campaign supporting the planting of trees and saving of forests. Write for a complete list of their materials which includes:

- **an Action Guide:** a 15-page booklet with basic information on global warming, a list of state coordinators and suggested activities on how to increase the number of trees in your area.

- **an Arbor Day Kit:** containing instructions for press conferences and ideas for community awareness.

- **a Curriculum Mini-Guide:** that has lesson plans for schools or service groups about global warming.

There is a small fee for materials. Write: Global Releaf, Dept. BH, P.O. Box 2000, Washington, D.C. 20013.

WORLD FOOD DAY

World Food Day was created by the member nations of the UN Food and Agriculture Organization (FAO). Now observed in more than 140 countries, the day focuses attention on food and farm problems, and encourages the people of the world to become more directly involved in the search for solutions.

The U.S. observance on October 16 is supported by over 400 private voluntary organizations, and the U.S. Department of Agriculture, but the main effort comes from the more than 15,000 local organizers in 50 states. Some of the materials available through the National Committee for **World Food Day** are: a study paper on food and environment issues; study/action packets of curriculum materials for K-12, college, and adults; bulletin inserts; posters; songs; book lists; film/videotape lists.

For information and names of supporting organizations in your community contact: National Committee for World Food Day, 1001 22nd St. N.W., Washington, D.C. 20437, (202)653-2404.

Utilize Church-affiliated Education Programs

Most churches operate regular educational systems such as Sunday Schools, adult education classes, and religious instruction. Some churches have more extensive programs: elementary schools, high schools and colleges or universities. Educators can designate some time to focus on the environment.

Grade school is an ideal time to teach children respect for creatures of the Earth. Respect comes when children know and understand the laws of the natural world. In addition to curriculum material developed for the classroom, schools can encourage nature walks, natural science projects, and trips to museums, zoos, and nature centers. Balanced nutritious meals, home surveys of potential indoor pollutants, school-wide recycling programs, gardening projects, as well as the studying of local environmental problems are also important components for school programs.

On the high school level, students can take advantage of programs that place them in a more active role with the environment. As part of their science courses or extracurricular activities, they can monitor the ecosystems of streams and rivers. They can participate in beach clean ups or they can initiate community recycling programs. Local governments and conservation agencies often work with students on these projects. Hands-on experience is an important stepping stone in the creation of responsible Earth citizens. (See the Organizational Resources for specific programs.)

The student's relationship to the poor, to those who do not have access to the Earth's resources, is also important. Educators can encourage community service and teach courses on the history of Third World countries.

Colleges and universities can commit a greater portion of their resources to environmental studies. They can offer courses that study ways to alleviate hunger and environmental pollution. Whenever possible, courses should seek to raise the level of social consciousness of the students. Schools can also give time to research renewable energy and other environmentally benign systems.

Colleges and universities can make it a practice to work with and for the poor. Land grant colleges, which currently cater heavily to agribusinesses, can help average or low income citizens as mandated by law. In fact, the agricultural schools were set up to help small and middle income farmers, who comprised a sizable portion of the population in the 1800's.

Educational mission opportunities, where teachers, agricultural experts, cooperative organizers, engineers, librarians, health and social workers, and scientists have an opportunity to serve the poor, abound. If institutions make it a regular practice to send out experts to help others, all can benefit.

Adult education classes can be offered either on specific environmental problems or on the skills necessary for effecting change. Retirees often have interest in returning to school. Tap this invaluable resource. Invite them to your classes.

Introducing Children to the Wonders of Nature

How do you instill in children not only a factual awareness of our planet, but also an appreciation and respect for Mother Earth? Rachel Carson, an author and environmentalist, wrote a classic book called *A Sense of Wonder*. In it, she explains the simple yet profound ways you can instill appreciation and respect. Check your local library.

Joseph Bharat Cornell wrote another classic that works well for large and small groups of children of any age. In *Sharing Nature with Children*, this naturalist says that the only prerequisites are a "happy combination of setting and receptivity." In one game he has groups of partners (one of each whom is blindfolded) walk into a wooded area. The blindfolded child is brought to a particular tree which he or she must explore—its texture, size, shape, smell, etc. The blindfolded child's partner then brings the child back to the starting place, the blindfold is removed, and the child, looking at a forest must find his or her own tree. His book is filled with similar nature-awareness experiences.

Creating Sacred Spaces

You can work with other church leaders to form linkages between your church and churches within your denomination, local churches of other denominations, and between local community churches and other communities. These cooperative activities usually occur "top-down" (from the national leadership to rank-and-file) in ecumenical and inter-religious models. You can begin a "bottom-up" environmentalism with local churches and ultimately influence regional and national church leaders. The challenge is to create cooperative ventures with other churches that offer practical, down-to-Earth activities.

While a number of interfaith activities are possible, one excellent choice is the creation of sacred spaces. The word "spaces" is used to emphasize the multiplicity of locations that can be set aside for matters of the spirit. Any community is or can become a sacred space wherein residents find peace and God's presence. These places could be found in members' own homes, in public buildings (such as the church), in natural settings, and in a person's own heart (if places are too crowded for privacy and solitude).

God is everywhere and believers find God in a host of settings. The existence of sacred space in the home, in the woods and in the heart does not diminish the place of church. Instead, these spaces emphasize the need to extend the sense of the sacred into the entire environment. Everything is created by God and we show reverence in home or out-of-doors as much as in a church building. The sacred and hallowed space of "church" can be extended.

Home as Sacred Space

The phone rings. Privacy is invaded by soliciting or wrong numbers. The television is on an average of six hours a day. The radio blares out its message from another room upsetting those desiring quiet. The noise and clutter of home sweepers, mixers, record players, and microwaves permeate much of our living and resting space and can numb our creative thought.

Some household members tolerate more noise bombardment than others, but no one is left unaffected. The home can be noise polluted and its environmental harmony threatened. Yet few raise their voices to such abuse.

Maybe the local church can intercede. The church preaches respect for individuals, the family, and the home. Threats to home and family fit within the moral concern of the church and noise pollution is very much an environmental hazard. Interfaith cooperation is needed to alert congregations to this overlooked area of environmental concern.

The church's mission may include ensuring that the locality has regulations safeguarding peace and quiet so that believers have the space to find God. Church groups could work to limit airplane arrivals and departures, develop traffic barriers by planting vegetation or building fences (especially near places where the elderly and those most affected by noise reside), and advocate for ordinances on the use of lawn mowers, motorcycles and chain saws.

The noise originating from within a house may be harder both to control and to change through regulatory procedures. Churches could encourage change by advocating that their members create quiet space within their homes by modifying their lifestyles. A number of ways to create quiet exist: installing curtains or acoustical tiles; covering closet, room walls, and ceilings with egg cartons. Most, but not all, Americans have sufficient room to designate a quiet space, but some are not able to do so. For these people to find quiet space, it may be necessary to go outside the home.

Church as Quiet Space

Some churches are left open all the time. In urban areas some are unlocked during the daylight hours only. Other church buildings are sub-divided into sections, some that are closed off and others that are open as sacred meditation chapels. It is important that a public, sacred place be provided. No community is so small that it may omit the designation of publicly accessible quiet space. The church is often the best indoor, sacred space. If your church cannot offer such a space, search and find a church that will, and publicize it in the community. You can sponsor this space jointly with another church community or faith.

Some prefer quiet space to be silent and dark. Others like a little music and a little light. Within a specific church, there are different ideas about what constitutes a sacred space. Variation may be part of the solution. If your community has two churches offering reflection space, try to insure that spaces are different from each other in order to serve people with different needs. Making such differences public is a salutary interfaith activity.

Outdoor Sacred Space

Besides the private and public sacred spaces, eco-church members remind others that God is truly everywhere and can be appreciated in the splendor of a hidden valley or a grove of hemlock. In fact, they can tell you exactly where the "best" places in public or private lands are located. Such spots may be rock formations, springs, mountain tops, a section of lakeshore, inlets, islands, cliffs, promontories, caves, or coves. Just about any quiet space in a natural setting can be a sacred outdoor space.

It is important to discover and champion your own local sacred spaces. These discoveries give as much a sense of ownership as does re-creating the home environment into sacred space. To make sacred a home and an outdoor space is one and the same act. The church pronounces both as Good News for all the Earth is sacred.

You are both a discoverer and a creator of sacred space. To make the spot more comfortable, you may be able to modify it by a tree stand, a blind, or a shelter. Better yet, allow the pristine area to remain such and instead, spend time making a new quiet space out of land that has been desecrated (see the "Prayer for Healing the Earth" on page 36.) For example, a scarred hillside may be reclaimed by planting raspberry canes in such a fashion that quiet coves are made until the land has time to return to health. Or the brush from a badly timber-cut hillside may be removed and new trees planted in order to create a place of rest and reflection. A host of other examples are possible.

Sacred Space in the Heart

A final area of environmental development is within time. People who are shut-in, incarcerated, or hospitalized may not have the freedom to go outdoors or to find quiet corners within their residences. If you are someone's spiritual guide, you can emphasize that God is found in quiet time as well as quiet space. Encourage people to find a moment away from busyness to pause and reflect. All believers need to create quality environmental time as well as space.

The individual who reserves this sacred time and space is not "making" God present, but realizing that God *is* present. With this realization a person may enter into a closer union with the Creator, speak intimately in prayer, and find the peace that surrounds and pervades sacred space. Through sacred time, persons recognize that they are carriers of God. They can appreciate themselves as sacred, bodily vessels of the holy and thereby come to treat their bodies and the Earth itself with greater respect.

Going Political

Living in harmony with the environment requires making political decisions. The diversion of development funds from a solid waste landfill system to a recycling program and the enactment of bottle bills can reduce the need for dumping areas. Federal or state legislated efficiency labels on all energy-using appliances and other products can have dramatic effects on lowering consumption of fossil fuels. Requiring coal-burning electric power plants to use cleaner fuel can curb the effects of acid rain on forests and lakes. Diverting highway funds to bike trails may encourage biking and make roads safer for youths and adults.

Direct Lobbying

Make a visit to the state or federal capital to lobby. First, study your particular environmental issue as thoroughly as possible. You can target key legislators or their staffs. If you cannot travel, give these key people at least a phone call on important occasions. Initiate letter writing campaigns among your congregation. You may want to support national or statewide lobby groups that work on environmental issues. (See the *Organizational Resources* at the end of the book for a list of agencies.)

Local, State and National Appointments

If the opportunity arises to serve on a governmental committee, give it prayerful consideration. It is the chance to get the environmental word into the competitive world of governmental affairs. Since the environment is only one of many issues facing elected officials, your ability to publicize these issues will tax your creativity. You may accept such an appointment and find the "honor" soon transforms into minute details, lengthy discussions, and heated arguments. Be willing to endure these if talent and time permit. You will have to be selective yet dedicated. The opportunity to educate others and initiate change exists.

Party Participation

Some people tend to avoid partisan politics. Others accept the limitations of political parties because they see the parties as vehicles of effective change. The movement towards the Greens, the environmental political parties of Europe, has not taken hold in the United States or Canada. Decide if you can have the greatest impact within or outside of traditional party structures. If you wish to go the first route, join and actively participate in the party that most reflects your political convictions. If you choose the second path be willing to develop political structures from the ground up. Designing structures of organization requires patience and hard work.

Political Office

Start on the bottom of the volunteer ladder and help prepare mailings. You can make phone calls, especially just before the elections, and attend rallies for the person you support. You as a backer may be able to make comments, give brief talks, shake hands with officials, and assist the candidate at key times. With time, experience and persuasion you may be moved to run for office. If so, place those knowledgeable about the environment on your advisory committee.

It must be noted that many church people conceive of themselves as *above* the path of direct politics and refrain from seeking public office. Encourage or become encouraged to see eco- politics as a natural unfolding of the power of the Eco-Church.

Testimony and Comments

Legislative action at local, state and national levels requires citizen input at a number of stages. Quite often environmental groups are overloaded and will welcome your expertise. Do not let the pecking orders of established groups dissuade you; there are always groups willing to sponsor your testimony. For example, if you wish to convert an unused railroad strip into a path, seek sponsorship from hiking groups or tourist associations. You can go it alone, but institutional identification is often helpful because legislators like the power of numbers and the implied voting influence.

Voting Registration

Churchgoers have a duty to vote and help elect environmentally responsible people to government. Encourage your church members to study the records and to participate in voter registration drives in your community. Take others to the polls on election day and assist in getting out the vote.

Oversight and Regulations

Enforcement of existing legislation and associated rule-making following enacted legislation may be more important than creating new laws. Study the laws and proposed regulations and talk with experts. Attending the public hearings is essential for communicating community concerns to enforcement officials. Write letters to newspapers if regulations are not being enforced. Visit the enforcers and let them know you are watching them. Remember that like other issues, environmental concerns can be "lost between the cracks."

Targeting Funds and Resources

Environmental causes often lack funds, resources and personnel. Recently church organizations and religious communities have started to offer loans or grants to environmental groups as part of their responsible social investment. If you are considering this type of investment, the following are pertinent questions to ask fundees:

- Does the group have an endurance record and is it on sound financial footing?

- Is it working on effective issues?

- Does the group perform environmentally benign operations?

- Are the people of the group committed to bettering the environment?

- Do they work well with other groups?

- Do they publicize their results?

One candidate for alternative investments might be a renewable-energy cooperative. Such a group could design and build solar heating systems. Here both investor and investee benefit: the investees develop self-sufficiency because they avoid steep interest payments and the investor learns of and becomes involved in the activities of the investees — and in turn learns about saving the Earth and about the need to be environmentally benign in activities. In this type of investment, networks are strengthened; long-term investments are encouraged; social change aspirations are shared. Finally, simple lifestyle acquired will reduce food, fuel, and health bills.

Although less successful, a stockholder's challenge is another political strategy. It is most successful when the stockholder resolution requests information or proposes a change in policy at an annual meeting. These challenges make a statement. If a change does not occur, churches must assess whether their continued investment in environmentally insensitive companies is in harmony with their spiritual values. Besides redirecting financial resources, your church can offer meeting space and supplies to environmental groups. You can also initiate a network among the different environmental groups and organizations within your community.

> ## A Model Program
>
> *Bread for the World* is a nationwide Christian movement that seeks justice for the world's hungry people by lobbying our nation's decision makers. This organization has an effective means of social action based primarily on letter- writing campaigns to elected officials.
>
> *Bread for the World* offers kits specifically designed to help Church leaders organize group participation. All members receive monthly newsletters explaining key issues, relevant bills and legislators to contact. Every year, the organization also targets one specific concern for a campaign called Offering of Letters. Guidelines for writing editorials or letters to the editor are also available.
>
> For more information, contact: Bread for the World, 802 Rhode Island Ave. N. E. Washington D.C. 20018 (202)-269-0200.

Bringing Hope through Alternatives

Suffering can be seen through hope-filled glasses. Repairing a broken people and Earth is part of the life-giving resurrection mystery. A spiritual power transcends physical and political powerlessness. God gives us a new heart (Ezek 36:27), a new spirit, a new way of acting. If we practice solidarity with the poor, we will seek victory using the meager resources the poor possess and we will truly empower them. We commit ourselves to small victories as preludes to larger ones.

Start a Community Garden

These gardens can be started in a variety of places by student groups, homeless shelters, soup kitchens, and nursing homes. The Green Guerrillas in New York City lists several steps to starting a garden:

- **Find a lot in your area.** Does it get direct sunlight? If so, how much? Are there any weeds or trees? What is the consistency of the soil? You may have to do a soil analysis. You will need written permission to develop a community garden.

- **Solicit financial support** from local merchants, community groups, or campus sources to cover insurance and expenses.

- **Plan and publicize** a general community meeting.

- **At the initial meeting get to know everyone,** detail ownership, size of plot(s), and background information (history of previous use for gardening or other activities). Encourage everyone consider what they want in the garden (picnic tables, trees, etc.) and set up a design committee. Decide on a name and install a sign with information about the way to contact the persons involved. Arrange for pick up of garbage and rubble.

Get Involved in Third World Issues

Ask friends from other countries what practices the U.S. has that damage their environment. You can cooperate with them in making changes so that these practices can be discouraged or halted. Justice groups working on Third World issues can be contacted. Those groups that assume environmental issues are included in the *Organizational Resources*. Some other things you can do to invite involvement in Third World issues:

- Display and sell materials from other countries.

- Sponsor a Third World parish, school or family.

- Encourage sermons on global awareness needs.

- Prepare Third World posters and displays for your church building.

- Make the needs of Third World people known to your congressional representatives.

Share Appropriate Technology

"Appropriate technology" is a concept derived from the insights of British economist E.F. Schumacher. It is a technology that is simple and it is the most environmentally and economically suitable technology for a given task. By definition it is accessible to low-income people. Low-income people usually cannot afford nor do they have the expertise to operate the more labor-saving, complex technologies.

Your local community may have people gifted in a form of appropriate technology; someone may know how to build a low-cost house, a wind mill, a compost toilet, a solar food dryer, a methane generator, or operate an organic garden. Encourage your members to share their expertise with others in need.

You can also promote groups like *Habitat for Humanity*, a group that utilizes vacationers and summer volunteers for rehabilitation projects and for construction of homes for low-income families.

Become a Peace Advocate

War and violence destroy our planet as well as take lives. You can encourage peace by joining with other church groups which specialize in resolving conflicts. Encourage those in the peace movement to see or consider environmental alternatives such as appropriate technology (solar heating, organic gardening, etc.) as peacemaking techniques. By making peace with parts of the Earth, we extend peace to all of it. It is a way to say that we want military defense transformed to eco-defense.

You might establish a nuclear-free zones. Certain areas of North America are now being declared "nuclear-free" of all nuclear weapons and warfare activities. Numerous counties, towns, townships, and cities have declared themselves to be nuclear-free. Besides the consciousness-raising potential, the practice is a start at addressing local community violence and is a form of environmentalism.

Respect for life also means extending respect to all parts of the life cycle from conception to death. An area of deep concern today is the reintroduction of the death penalty by a number of states that had been under an imposed moratorium awaiting Supreme Court decisions. Speak out against the death penalty.

In essence, transforming the peace movement to one of saving the planet is a must for the Eco-Church.

Consider what children are being fed by television and the toy industry. Express your views to television stations. Speak out against toy war weapons. Write to War Resisters League, Box 1093, Norwich, CT, 06360 for its "Stop War Toys Campaign" packet.

One Success Story: A Process

The efforts of the school children from Salt Lake City mentioned in the beginning of this chapter illustrate how idealism and realism work hand-in-hand. The following is a summary of their plan of action as recorded in the March 1989 issue of *Sierra Magazine*.

- identified over 50,000 rusted barrels stockpiled for over 40 years in an unprotected area.
- queried Health Department to see if water was contaminated. Health Department responds with brush-offs and delays.
- door-to-door survey of neighborhood pinpoints well water for testing and informs residents of potential problems. Community unresponsive but some TV and newspaper coverage.
- approached owner for information.
- researched hazardous waste in major journals and through periodical clipping service.
- invited environmental consultants, health officials and Salt Lake City Emergency Hazardous Waste Clean Up Team to lecture to class.
- contacted EPA national hotline and previous owner of yard.
- visited Mayor's office who takes action and promises clean up within 18 months.
- local TV coverage and Denver EPA gets involved in testing.
- raised almost $500 for clean up but a law restricts their contribution to this cause.
- EPA test results confirmed toxic substances contaminating soil and water.
- barrels removed.
- students made presentations to community groups.
- children initiated state law to parallel National Superfund that helps clean up of abandoned toxic-waste sites and allows their contributions to be used at their site.
- A year and half after the class project began, the bill is passed.

Earth Commitment

I resolve:

to honor and love the Earth
as mother to us all;

to always speak with respect about the
produce of the Earth as though plants, animals
and human beings belong to one family;

to have contact with the Earth through
growing part of my food from the soil;

to understand the fragility of the Earth and
appreciate how easily it can be damaged;

to be sensitive to the way the I can harm
the Earth through exploitation or pollution;

to live lightly on the Earth;

to clean up litter where I find it;

to show compassion for the parts of the Earth
that suffer most from greed and carelessness;

to resist in a non-violent manner the aggressors
who seek to ruin the Earth and those who despair
of finding any future for the Earth or its creatures;

to halt the extinction of any plant or animal species;

to assist those who do not share the Earth's produce
so that all may have the basics of human life;

to work with others in renewing the Earth and
announcing a time of peace and prosperity;

to see the Earth as God's garden in which
all creatures share the gift of life;

and to pray for the Earth and its creatures.

Environmental Resources

Educational Resources

PROJECT LEARNING TREE

An award-winning environmental education program designed for teachers and other educators working with students in kindergarten through grade 12. PLT uses the forest as a "window" into the natural world, helping young people gain an awareness and knowledge of the world around them, as well as their place within it. Over 175 activities help teach science, mathematics, language arts, social studies, humanities, and other subjects. Appropriate for urban, suburban and rural areas. PLT resources and activities are taught at a six-hour workshop held in various locations through out the year. Materials available at the workshop.

Contact: Project Learning Tree, 1250 Connecticut Ave. N.W. Suite 320, Washington D.C. 20036, (202) 463-2472.

PROJECT WILD

Another award-winning environmental and conservation education program for kindergarten through grade 12. Appropriate for classroom use, community education programs with youth, church and scouting programs, and in outdoor education. Three guides with instructional and interdisciplinary activities address such topics as awareness and appreciation of wildlife, human values and wildlife, ecological systems, conservation and responsible human actions. Activity Guides are available free of charge to participants in Project Wild workshops. Workshops typically also are provided free of charge.

Contact: Project Wild, P.O. Box 18060, Boulder, CO 80308-8060, (303) 444-2390.

ADOPT-A-STREAM

Adopt-a-Stream is a model program headed by Dr. Betsy Brauer of Rochester, New York that has expanded to a dozen states. It offers programs to high school students who are interested in testing and monitoring their own water. Consider applying an adaptation of this program to forest preservation, wildlife protection, air quality, hazardous waste dumping, dam deterioration or road protection.

Contact: Adopt-a-Stream, Delta Laboratories, Inc. 34 Elton Street, Rochester, New York 14607 (716) 271-5333.

NATIONAL AUDUBON SOCIETY	*EXPEDITION INSTITUTE:* For high school, college and graduate courses and summer expeditions which focus on environmental awareness and action. Contact: Northeast Audubon Center, Sharon, Connecticut 06069.
SPIRITUALITY AND THE EARTH	A curriculum for adult and senior high school groups. Contact: Connections, Inc. 416 Sixth Street, Traverse City, MI, 49684.
TODAY'S WORD FOR ADULTS	Level 9: *Living the Word,* "To The Earth." A 13-session course in practical Christian environmentalism. Published quarterly. Contact: The Christian Board of Publication, 1316 Convention Plaza, P.O. Box 179, St. Louis, MO, 63166.
BREAK-THROUGH ON HUNGER	VHS videotape and study guides based on a four-part PBS series on the ethical and spiritual dimensions of world hunger. Suggested study process is designed to clarify confusing connections between our domestic priorities and those of our global neighbors. It also encourages participants to apply their religious convictions to this topic. The Interreligious Coalition for Breakthrough on Hunger was formed by Protestant, Roman Catholic, and Jewish groups in order to share their viewpoints with the TV producers. Contact: Alternative Coalition. P.O. Box 429, 52 63 Boulder Crest Rd, Ellenwood, Georgia 30049. The Interreligious Coalition for Breakthrough on Hunger Headquarters office is 475 Riverside Dr., 16th Floor, New York, NY 10115. (212)-870-2951.
WORLD SERVICE FILM LIBRARY	Audio-visuals available on a free loan basis. Included are *Choices: Positive Alternative to Solid Waste Disposal; Keepers of the Forests,* a description of the destruction of the rain forests and the problems of a group of indigenous people; and *The Lorax* a Dr. Seuss tale about taking natural resources for granted. Contact: World Service Film Library, P.O. Box 968 28606 Phillips St. Elkhart, IA 46515 (219) 264-3102

BEFRIENDING CREATION	*Befriending Creation Newsletter.* A newsletter to raise peace toward the environment.

Contact: Jack Phillips and Robert Shutz, Editors, 7899 St. Helena Rd., Santa Rosa, CA. 95404. |
| DISCOVER AND SHARE | "Discover and Share," is a small group of faith and life sharing men and women who meet monthly. They come together to share and support one another and to exchange on social, economic and ecological justice issues.

Contact: Bev Koch, 6822 N. Interstate, Portland, OR 97217. |
| ECHO | ECHO, Inc. is a Christian, non-profit, interdenominational organization based on a five-acre farm in Southern Florida. They publish *ECHO Development Notes* and "Academic Opportunity Sheets" for the Third World laboratories and libraries.

Contact: ECHO, Inc., R.R. 2 Box 852, North, Fort Myers, FL 33903, (813) 997-4713. |

Organizational Resources

1. **Acid Rain Foundation**, 1410 Varsity Dr. Raleigh, NC 27606, (919) 828-9443.

    Seeks to educate people about acid rain through conferences and through a resource library, open to the public.

2. **Alternative Energy Association**, 366 Woodknoll Terrace, Wyoming, OH 45215.

    Local clearinghouse organization initiating numerous projects in solar and thermal home creation.

3. **American Council for an Efficient Economy**, 1001 Connecticut Ave., N.W., Suite 535, Washington, DC 20036, (202) 429-8873.

    An energy conservation organization.

4. **American Farmland Trust**, 1920 N St. N.W., Suite 400, Washington, DC 20036, (202) 659-5170.

    Seeks to preserve agricultural land, soil, and farming opportunities through inquiry, consulting, referrals, and seminars.

5. **American Lung Association**, 1740 Broadway Ave., New York, NY 10019, (212) 315-8700.

    Teaches preventative measures against lung disease, and assists individuals with breathing difficulties to acquire the proper care and to cope with the lung problems.

6. **American Public Health Association**, 1015 Fifteenth St. N.W., Washington, DC 20005, (202) 789-5674.

    Studies public health through dental health, environment, epidemiology, nutrition, and gerontological health. Offers consulting and referrals.

7. **Americans for Safe Food**, 1875 Connecticut Ave. N.W., Suite 300, Washington, DC 20009, (202) 332-9110.

    Works to ensure good food practices.

8. **Appalachia—Science in the Public Interest**, P.O. Box 298, Livingston, KY 40445, (606) 453-2105.

    Ecological research and demonstration facility makes science and technology responsive to the needs of low-income Appalachians.

9. **Arts for the Environment**, Mark Spencer, Bascom Lodge, P.O. Box 1652, Lanesborough, MA 01237, (413) 743-1591.

    Environmentally sensitive graphic design, illustration, and layout.

10. **Association of State Drinking Water Administrators**, 1911 N. Fort Meyers Dr., Suite 400, Arlington, VA 22209, (703) 524-2428.

11. **Atlantic Center for the Environment**, QLF Program Headquarters, 39 South Main St., Ipswich, MA 01938-2321, (508) 356-0038.

    Teaches ecology and environmental preservation through research, workshops, courses, publications, and references.

12. **Au Sable Institute**, 2508 Lalor Rd., Oregon, WI 53575, (608) 222-1139.

   A practical education program in environmental ethics with an evangelical Christian orientation.

13. **BioIntegral Resource Center**, P.O. Box 7414, Berkeley, CA 94707, (415) 524-2567.

   Researches and provides information on Integrated Pest Management and other methods of less toxic pest control.

14. **BioRegional Project**, David Haenke, Rt. 1 Box 20, Newburg, MO 65550, (314) 762-3423.

   Philosophy of and general information on bioregionalism.

15. **The Catalyst Group**, 139 Main St. Suite 614, Battleboro, VT 05301, (802) 254-8144.

   A social investment information group.

16. **Center for Economic Conversion**, 222 C View St., Mountain View, CA 94041, (415) 968-8798.

   Provides public education on issues of peace and defense conversion.

17. **Center for Holistic Resource Management**, P.O. Box 7128, Albuquerque, NM 87194, (800) 654-3619 FAX: (505) 247-1008.

   Formed by a group of ranchers, farmers, researchers and environmentalists to serve as a focal point for the exchange and dissemination of knowledge on holistic management.

18. **Center for Law and Social Policy**, 1751 N. St. N.W., Washington, DC 20036, (202) 328-5140.

   Legal education organization for problems with the poor and minorities, civil rights enforcement and health care advocacy.

19. **Center for Maximum Potential Building Systems, Inc.**, 8604 F.M.969, Austin, TX 78724, (512) 928-4786.

   Educational research and development organization working on appropriate technology systems.

20. **Center for Rural Affairs**, P.O. Box 405, Walthill, NE 68067, (402) 846-5428.

   Center for agricultural research and social and ecological impact studies. Focus is toward Nebraska residents.

21. **Center for Science in the Public Interest**, 1501 16th St. N.W., Suite 928, Washington, DC 20036, (202) 332-9110.

   Seeks to improve the American diet through research and education. It provides advisory services and publications sold at cost.

22. **Chesapeake Bay Foundation**, 162 Prince George St., Annapolis, MD 21401, (301) 268-8816.

   Seeks citizen involvement in the environment of the Chesapeake Bay Region. Provides advisory, referrals, and library services.

23. **Children's Foundation**, 725 15th St. N.W., Suite 505, Washington, DC 20005, (202) 347-3300.

   Monitors the administration of programs for families and children, especially child care and food programs.

24. **"Citizen Alert"**, Box 1681, Las Vegas, NV 89125, (702) 648-8982.

   Monitors and alerts the public on hazardous waste. Especially concerned with nuclear waste from underground testing.

25. **Citizens Clearinghouse for Hazardous Waste**, P.O. Box 926, Arlington, VA 22216, (703) 276-7070.

> Group offering assistance with hazardous waste problems through consulting, referrals, and publications.

26. **Clean Water Action Project,** 1320 18th St. N.W., Washington, DC 20036, (202) 457-1286.

> Distributes publications and offers citizen training in outreach programs concerning water and toxics.

27. **Coast Alliance**, 235 Pennsylvania Ave. S.E., Washington, DC 20003, (202) 546-9554.

> Addresses issues concerning population growth and its effect on the environment.

28. **Community Nutrition Institute (CNI)**, 2001 S St. N.W., Washington, DC 20009, (202) 462-4700.

> Seeks to inform community and consumer groups and local federal food officials of the need for a national food policy. Offers consulting, seminars, and publications.

29. **Concern, Inc.**, 1794 Columbia Rd. N.W., Washington, DC 20009, (202) 328-8160.

> Offers environmental information and encourages community action over ground water, pollution, pesticides, acid rain, and toxic waste disposal.

30. **Conservation Renewable Energy**, P.O. Box 8900, Silver Spring, MD 20907, (800) 523-2929.

> An alternative energy information center.

31. **Council of State Governments**, P.O. Box 11910, Iron Works Pike, Lexington, KY 40578, (606) 252-2291.

> Gives technical assistance to state officials and legislators on meeting the health needs of the elderly and on recycling.

32. **Co-op America**, 2100 M St.,N.W. Suite 403, Washington, DC 20063, (800) 424-2667.

> Facilitates the networking of nonprofit alternative organizations. Provides catalogs and quarterly publication.

33. **Council on Economic Priorities**, 30 Irving Place, New York, NY 10003, (212) 420-1133.

> Surveys corporations about social concerns including military production, consumerism, hiring policies, pollution control, foreign investments, and product safety.

34. *Creation,* P.O. Box 19216, Oakland, CA 94619, (415) 253-1192.

> An magazine containing Earth and spirituality topics.

35. **Critical Mass Energy Project**, 215 Pennsylvania Ave. S.E., Washington, DC 20003, (202) 546-4996.

> Division of Public Citizen that researches and reports on nuclear and alternative energy issues. Publishes reports and responds to public inquiry.

36. **CROP**, P.O. Box 968, 28606 Phillips St., Elkart, IN 46515, (219) 264-3102.

> Provides worldwide local relief services in conjunction with Church World Service.

37. **Defense Budget Project**, Center on Budget and Military Priorities, 236 Massachusetts Ave. N.E., Suite 401, Washington, DC 20002, (202) 408-1517.

   Researches military expenditures in relation to total budget with an aim toward reallocation of these funds for social programs.

38. **Delaware Valley Toxics Coalition**, 125 South Ninth St., Suite 700, Philadelphia, PA 19107, (215) 627-7100.

   Pesticides.

39. **E.F. Schumacher Society**, Box 76, RD 3, Great Barrington, MA 01230, (413) 528-1737.

   Supporters of the "Small is Beautiful" perspective, providing social and appropriate technology resources.

40. **Environmental Action Foundation**, 1525 New Hampshire Ave. N.W., Washington, DC 20036, (202) 745-4871.

   A lobbying organization that works to elect environmental candidates to congress and sends notices to its members. Includes Energy Conservation Coalition.

41. **Environmental Defense Fund**, 257 Park Avenue S., New York, NY 10010, (212) 505-2100.

   Advocates alternative practices to the abuses of energy, toxic materials, water, and wildlife.

42. **Environmental Protection Agency Office of Drinking Water**, 401 M St. S.W., WH-550E, Washington, DC 20460, (202) 382-5522.

43. **Food First**, Institute for Food and Development Policy, 145 Ninth St., San Francisco, CA 94103, (415) 864-8555.

   Public education group focusing on hunger and world development issues.

44. **Food Irradiation Response**, Box 5183, Santa Cruz, CA 95063, (408) 684-0577.

   Advocacy group for food safety.

45. **Food Safety and Inspection Service USDA**, 14th and Independence Ave. S.W., Washington, DC 20250, (202) 447-6313.

   Does mandatory inspection of meat and poultry. Offers information and referrals.

46. **Florida Atlantic University Joint Center for Environmental and Urban Problems**, 220 Southeast Second Ave., Fort Lauderdale, FL 33301, (305) 355-5270.

47. **Fourth World Review**, 24 Abercon Place, London, NW8, England.

   Raising consciousness of church groups to the problems of the world's most destitute nations.

48. **Friends of the Earth**, 218 D St., S.E., Washington, DC 20003, (202) 544-2600.

   A national environmental and energy information center offering seminars, consultation, and publication listings.

49. **Funding Exchange**, 666 Broadway, 5th Floor, New York, NY 10012, (212) 529-5300.

   Social investment forum.

50. **GAIA Studios**, 327 Dempster #2E, Evanston, IL 60201, (312) 475-1553.

   Artistic creations with the Earth in mind.

51. *GARBAGE, The Practical Journal for the Environment,* Old House Journal Corp., 435 Ninth St., Brooklyn, NY 11215, (718) 788-1700.

> Committed to addressing the questions people are asking about waste disposal, landfills, dirty air, and toxics, as well as indoor hazards and food and water supply hazards.

52. **Genesis Farm**, Box 622, Blairstown, NJ 07825, (201) 362-6735.

> Practical permacultural and model Earth spirituality farm.

53. **Greenlight**, 4014C Utah Avenue, Nashville, TN 37209.

> Cumberland Green Bioregional Council newsletter includes listing of events in the Bioregion.

54. **Greenpeace USA**, 1436 "U" St. N.W., Washington, DC 20009, (202) 462-1177.

> Environmental activist network which lobbies the government and stewards the oceans, monitoring and blocking commercial and government shipping practices that are environmentally destructive, including those that kill dolphins and whales.

55. **Health Research Group–Public Citizen**, 2000 P St. N.W., Suite 700, Washington, DC 20036, (202) 872-0320.

> An advocacy group working on a variety of consumer health issues.

56. **Heifer Project International**, P.O. Box 808, Little Rock, AR 72203, (501) 376-6836.

> Develops and operates small farm and livestock programs to alleviate hunger.

57. **Hort Ideas**, Rt. 1, Box 302, Gravel Switch, Ky 40328.

> Quarterly newsletter condensing numerous discoveries, techniques, pieces of advice, products, books, and opportunities for gardeners found in the stream of current horticultural publications.

58. **Illinois South Project, Inc.**, 116 1/2 W. Cherry, Herrin, IL 62948, (618) 942-6613.

> Coal mining regulation and other regional issues.

59. **IMAGO**, 553 Enright Ave., Cincinnati, OH 45205, (513) 921-5124.

> A grassroots organization emphasizing Earth's sacredness through celebrations, educational programs, and the development of an interdependent urban neighborhood. Has a bi-monthly newsletter.

60. *Indoor Air News,* Consumer Federation of America, 1424 16th St. N.W., Washington, DC 20036.

> Periodical reviewing indoor products and their safety from hazardous emissions.

61. **Institute for Alternative Agriculture**, 9200 Edmonston Rd, Suite 117, Greenbelt, MD 20770, (301) 441-8777.

> A research center devoted to assisting people in the development of a regenerative, low input agriculture. Offers a newsletter, conferences, and use of its resource collection.

62. **Institute for Earth Education,** Box 288, Warrenville, IL 60555, (312) 393-3096.

> Foundation seeking to build in people of all ages an understanding of, an appreciation for, and a harmonious relationship with the earth.

63. **Institute for Local Self Reliance,** 2425 18th St. N.W., Washington, DC 20009, (202) 232-4108.

> Encourages cities to use self reliance as a development strategy providing consulting, referrals, seminars and publications.

64. **Interfaith Center on Corporate Responsibility,** 475 Riverside Dr., Room 566, New York, NY 10115, (212) 870-2936.

> Coordinates church proxy resolution activity offering referrals and publications.

65. **International Alliance for Sustainable Agriculture,** 1701 University Ave. Minneapolis, MN 55414, (612) 331-1099.

> Networks and forms conferences for ecological growers worldwide.

66. **International Network for Religion and Animals,** 2913 Woodstock Ave., Silver Spring, MD 20910, (301) 565-9132.

> Worldwide Animal Protection Agency with a theological basis.

67. **International Society for Environmental Education:** 1. Dept. of Planning, U. of Oregon, Eugene, OR 87403, (503) 346-3895. 2. *The Biosphere*, School of Natural Resources, Ohio State University, 2021 Coffey Rd., Columbus, OH 43210.

68. **Investor Responsibility Research Center,** 1755 Massachusetts Ave., N.W., Washington, DC 20036, (202) 234-7500.

> Researches and publishes newsletters on corporate responsibility. Gives referrals and information to subscribers.

69. **Izaak Walton League of America,** 1401 Wilson Boulevard, Level B, Arlington, VA 22209, (703) 528-1818.

> Organization seeking to mobilize citizens on environmental issues. Offers Publications, Referrals, and Conferences.

70. *KATUAH JOURNAL* P.O. Box 638, Leicester, NC 28748, (704) 754-6097.

> A quarterly ecology journal for the Southern Appalachian Bioregion, networks events and movements, depicts mountain culture, and includes essays, poetry, and Native American ritual.

71. **Kerr Center for Sustainable Agriculture,** Highway 271-S, Box 588, Poteau, OK 74953, (918) 647-9123.

> Demonstrates sustainable ranching and researches diversified farming systems incorporating aquaculture, horticulture, poultry and covercrops.

72. **Land and Liberty,** 177 Vauxhall Bridge Rd., London, S.W.I., Great Britain.

> A British bi-monthly journal on the history and current issues of land reform, labor, and taxes. Covers key British and international figures affecting land rights.

73. **Land Institute,** 2440 East Water Well Rd., Salina, KS 67401, (913) 823-5376.

> A research, education facility devoted to restoring the Prairie, and developing perennial grain food crops with native grasses through no-till methods.

74. **Land Stewardship Council,** P.O. Box 25719 Raleigh, NC 27611, (919) 836-1990.

   Working on environmental land use and related issues.

75. **League of Women Voters,** 1730 M St., SE, Washington, DC 20036, (202) 429-1965.

   Aids people in right-to-know issues and in working through the correct legislative avenues for successful change-making in social and environmental concerns.

76. **Long Branch Environmental Education Center,** Route 2, Box 132, Leicester, NC 28748, (704) 683-3662.

   Demonstration Center with programs in Aquaculture, organic gardening, tree saving, and local waste disposal outreach.

77. **Louisiana Toxics Project,** 3227 Canal St., New Orleans, LA 70119, (504) 482-9566.

   Toxic substances and water quality issues.

78. **Meadowcreek Project,** Fox, AR 72051, (501) 363-4500.

   An environmental organization offering demonstration in sustainable agriculture, passive and active solar systems, and resource conserving manufacturing.

79. **National Audubon Society,** 801 Pennsylvania Ave. S.E., Washington, DC 20003, (202) 547-9009.

   Organization existing in honor and protection of wildlife ecosystems.

80. **National Association of Conservation Districts,** 509 Capital Court N.E., Washington, DC 20002, (202) 547-6223.

   Soil conservation.

81. **National Association of State Foresters,** 444 North Capitol St. N.W., Washington, DC 20001, (202) 624-5415.

82. **National Campaign Against Toxic Hazards,** 1168 Commonwealth, Boston, MA 02134, (617) 232-0327.

83. **National Center for Appropriate Technology,** 3040 Continental Dr., P.O. Box 3838, Butte, MT 59702, (406) 494-4572.

   Develops appropriate technologies. Works with community leaders and policy makers to help communities implement alternative energy systems.

84. **National Center for Policy Alternatives,** 1875 Connecticut Ave. N.W., Suite 710, Washington, DC 20009, (202) 387-6030.

   Resource and activation center for environmental policy and research.

85. **National Clean Air Coalition,** 801 Pennsylvania Ave. S.E., Washington, DC 20003, (202) 624-9393.

   A lobbying group for a strong national clean air policy, provides resources for researchers and gives conferences.

86. **National Coalition Against the Misuse of Pesticides,** 701 E St. S.E., Suite 200, Washington, DC 20003, (202) 543-5450.

   Community based group working for pesticide safety and the use of alternative methods of pest management.

87. **National Conference of State Legislatures,** 1560 Broadway, Suite 600, Denver, CO 80202, (303) 623-7800.

   Drinking water issues.

88. **National Peace Academy Campaign**, 110 Maryland Ave. N.E., Suite 409, Washington, DC 20002, (202) 546-9500.

Peace, defense conversion.

89. **National Gardening Association**, 180 Flynn Ave., Burlington, VT 05401, (802) 863-1308.

Publishes a newsletter and several books on creating community, employee, prison, and youth gardens. Main goal is to encourage people of all walks and in all places to grow gardens for their own produce.

90. **National Pesticide Telecommunications Network,** Texas Tech University, Science Health Center, Fourth and Indiana 1A111, Lubbock, TX 79430, (800) 858-7378.

Provides pesticide lab services and info to prevent pesticide accidents. Has extensive library.

91. **National Small Flow Clearing House**, West Virginia University, P.O. Box 6064, Morgantown, WV 26506-6064, (800) 624-8301.

Information on issues of water quality including water purification and waste water treatment.

92. **National Water Summary**, USGS, 12201 Sunrise Valley Dr., MS 407 Reston, VA 22092, (703) 648-6856.

Water pollution.

93. **National Wildlife Federation**, 1400 16th St. N.W., Washington, DC 20036, (202) 797-6800 or (800) 432-6564.

Publishes books on wildlife and Natural Resources. Offers references and services and use of its library at its Virginia branch office.

94. **Natural Resources Defense Council**, 40 West 20th St. New York, NY 10168, (212) 727-2700 or 1350 New York Ave., NW, Washington, DC 20005, (202) 783-7800.

Provides advisory services for issues involving environmental law.

95. **New Alchemy Institute**, 237 Hatchville Rd., East Falmouth, MA 02536, (508) 564-6301.

An international non-profit, publishing, organization combining several forms of appropriate technology in agriculture, aquaculture, and solar design for an eco-living model.

96. **New Creation Institute**, 518 South Ave. W., Missoula, MT 59801, (406) 721-6704.

An educational program in environmental ethics.

97. **New Jersey Coalition for Alternatives to Pesticides**, P.O. Box 627, Boonton, NJ 07005, (201) 334-7975.

Researches safe substitutes for agricultural chemicals.

98. **Northern Plains Resources Council**, Rm 419 Stapleton Building, Billings MT 59101, (406) 248-1154.

Lobbies for policies in coal energy, agriculture, and water resources for the Northern Plains. Offers resource library on coal developments.

99. **Nuclear Information and Resources Service**, 1424 16th St., N.W., Suite 601, Washington, DC 20036, (202) 328-0002.

Provides information on nuclear energy including a bi-monthly magazine, and publications.

100. **Office of Regulatory Affairs**, Food and Drug Administration,. 5600 Fishers Lane, Rockville, MD 20857, (301) 443-6200.

101. **Oregon Department of Environmental Quality**, 811 6th Ave., S.W., Portland, OR 97204, (503) 229-6046.

Resource for practical recycling solutions.

102. **Physicians for Social Responsibility**, 1601 Connecticut Ave. N.W., Suite 708, Washington, DC 20009, (202) 785-5371.

Nuclear issues.

103. **Planet Drum Foundation**, P.O. Box 31251, San Francisco, CA 94131, (415) 285-6556.

A Bioregional foundation.

104. **Population Reference Bureau, Inc.**, 1875 Connecticut Ave., N.W., Suite 520, Washington, DC 20009, (202) 483-1100.

Ecological education foundation. Provides information on population trends. Publishes data and offers consulting for a fee.

105. **Rachel Carson Council**, 8940 Jones Mill Rd., Chevy Chase, MD 20815, (301) 652-1877.

International Environmental Clearinghouse with special interest in chemicals, especially pesticides. Provides seminars, referrals, and publications.

106. **Radioactive Waste Campaign**, 7 West St. New York, NY 10090, (914) 986-1115.

Sells informational materials dealing with nuclear waste issues.

107. **Rainforest Action Network**, 301 Broadway Suite A, San Francisco, CA 94133, (415) 398-4404.

An international activist and lobbying organization that issues a newsletter that includes current rainforest destruction zones and who to write to lobby in protection.

108. **Renew America**, 1400 Sixteenth St. N.W., Suite 710, Washington, DC 20036, (202) 232-2252.

An environmental policy and research organization which runs the "Searching for Success" Program.

109. **Renewable Energy News Digest**, P.O. Box 295, Hurley, NY 12443. 110. **Resource Recycling**, P.O. Box 10540, Portland, OR 97210, (503) 227-1319.

111. **Resources for the Future**, 1616P St. N.W., Washington, DC 20036, (202) 328-5000.

Research organization concerned with conservation of natural resources. Has special documentation of the Ford Foundation Energy Policy Project.

112. **Rocky Mountain Institute**, 1739 Snowmass Creek Rd., Snowmass, CO 81654-9199, (303) 927-3851.

Research foundation seeking global security through efficient and sustainable use of energy. Integrative building design, hydrology, agriculture, and national security are some areas of work.

113. **Rodale-Press, Inc.**, Emmaus, PA 18049.

Produces *Organic Gardening* and *Prevention* magazines and several gardening, home construction, and health books.

114. **Rodale Research Center**, 611 Siegfried Dale Rd, Kutztown, PA 19530, (215) 683-6383.

> Experiments with alternative food crops and equipment to offer sustainable possibilities to the farming industry. Also extends research and service to the third world, working with people to create solutions in feeding their populations.

115. **Rural Southern Voice for Peace (RSVP)**, 1901 Hannah Branch Road, Burnsville, NC 28714, (704) 675-5933.

> Education, training and organization assistance for peace activists in the Southeast.

116. **Safe Energy Communication Council**, 1717 Massachusetts Ave., NW, Suite LL 215, Washington, DC 20036, (202) 4371-1000.

> An alternative energy networking and support organization.

117. *SEEDS*, 222 East Lake Dr., Decatur, GA 30030, (404) 378-3566.

> Magazine addressing issues of hunger, poverty, and food regeneration.

118. **Sierra Club**, 730 Polk St., San Francisco, CA 94108, (415) 776-2211.

> Environmental organization which organizes outings across the country and lobbies through legal avenues to protect the natural environment.

119. **Society of American Foresters**, 5400 Grosvenor Lane, Bethesda, MD 20814, (301) 897-8720.

> Represents all branches of the forestry profession. Referrals and resources available.

120. **Soil Conservation Service (USDA)**, P.O. Box 28901, South Agricultural Building, Washington, DC 20013, (202) 447-6267.

> Publishes material on soil and water conservation. Offers library and referrals.

121. **Soil Conservation Society of America**, 7515, N.E. Ankeny Rd., Ankeny, IA 50021, (515) 289-2331.

> Non-profit organization seeking to foster sound land use through soil and water conservation. Offers position statements, conferences and publications for sale.

122. **Southwest Research and Information Center**, P.O. Box 4524, Albuquerque, NM 87106, (505) 262-1862.

> Public advocacy group, professionally backing environmental and social cases through research, referrals, resources, and court testimony.

123. **Swedish NGO Secretariate on Acid Rain**, Miljovard, Box 245, S-401 24 Gotenborg, Sweden Phone: +46-31-15 39 55.

> International networker for a global response to acid rain problems.

124. **The TILTH Association** (206) 633-0451 (PINA), 4649 Sunnyside Ave. N., Seattle, WA 98103.

> Structures programs and produces newsletter, *The Permaculture Activist*, for ongoing landscape planting design ideas and opportunities (permanent culture).

125. **Tranet and Rain Magazine**, P.O. Box 567, Rangeley, ME 04970, (207) 864-2252.

> A network for appropriate/alternate technology including solar energy, wind, rural, and village technologies, citizen action, and future studies. Publishes and gives referrals.

126. **Union of Concerned Scientists,** 26 Church St., Cambridge, MA 02238, (617) 547-5552.

> Scientists and citizens concerned with nuclear issues, arms control and national energy policy. Participates in research, lobbying and educational programs.

127. **Urban Land Institute,** 625 Indiana Ave. N.W., Washington, DC 20004, Suite 400, (202) 624-7000.

> Provides publications and project files on urban land use and free referral for urban planning.

128. **Washington Environmental Council,** 5200 University Way N.E., Seattle, WA 98105, (206) 527-1599.

> Forest management issue.

129. **West Virginia Mountain Stream Monitors,** Box 170, Morgantown, WV 26505.

> Promotes voluntary stream monitoring in West Virginia. Much of their work involves reclamation of waterways damaged by strip mining.

130. *Wild Earth*, P.O. Box 492, Canton, NY, 13617.

> Radical environmental journal which incorporates "deep" ecological explorations with the fury of wilderness devotees fighting back against damage to the Earth.

131. **Wisconsin Rural Development Center,** 1406 Hwy 18-151 East, Mount Horeb, WI 53572, (608) 437-5971.

> Advocacy group focusing on agricultural issues and saving the family farm.

132. **Woodlands Mountain Institute,** Main and Dogwood Streets, P.O. Box 907, Franklin, WV 26807, (304) 358-2401.

> Educational and scientific organization that promotes the advancement of mountain cultures and preservation of mountain environments.

133. **Work on Waste, USA, Inc.,** 82 Judson Ave., Canton, NY 13617, (315) 379-9200.

> Assists organizations nation-wide in fighting incinerators and in dealing with other issues of waste management.

134. **World Peacemakers,** 2025 Massachusetts Ave. N.W., Washington, DC 20036, (202) 265-7582.

> Advocacy group focusing on world peace negotiations.

135. **World Wildlife Fund and Conservation Foundation,** 910 17th St. N.W., #619, Washington, DC 20006, (202) 293-4800.

> Provides newsletter, publications, research, policy studies and conferences on wildlife preservation. Encourages enhancement and protection of all life.

136. **Worldwatch Institute,** 1776 Massachusetts Ave. N.W., Washington, DC 20036, (202) 452-1999.

> Measures and reports emerging environmental, natural resource, and policy issues affecting human survival on a global scale. Publishes Worldwatch Magazine and State of the World Annual.

137. **Yurt Foundation,** Bucks Harbor, ME 04618.

> Combines traditional folk wisdom and modern appropriate technology to encourage simple living and a mutually fulfilling relationship with the Earth.

Books of Interest

Austin, Richard Cartwright. *Baptized into the Wilderness.* Atlanta: John Knox Press, 1987.

_____. *Beauty and the Lord.* Atlanta: John Knox Press, 1988.

Barbour, Ian G. *Earth Must Be Fair: Reflections On Ethics, Religion, And Ecology.* Englewood Cliffs, New Jersey: Prentice Hall, 1972.

Barnette, Henlee H. *The Church and the Ecological Crisis.* Grand Rapids, Michigan: Wm.B. Eerdmans, 1972.

Berry, Thomas. *The Dream of the Earth.* San Francisco: Sierra Club Books, 1988.

Berry, Wendell. "A Secular Pilgrimage," in *Western Man & Environmental Ethics.* Ed. Ian G. Barbour. Menlo Park, California: Addison-Wesley, 1973.

_____. *The Unsettling of America.* New York: Avon, 1977.

Birch, Bruce C. and Larry L. Rasmussen. *The Predicament of the Prosperous.* Philadelphia: The Westminster Press, 1978.

Birch, Charles and John B. Cobb, Jr. *The Liberation of Life From Cell to the Community.* Cambridge, England: Cambridge University Press, 1981.

Brown, Lester R., *State of the World.* New York: W.W. Norton. (An annual publication issued in the beginning of the year.)

Bryant, Audrey. "The Feminine Element," in *God's Green Earth.* London: Christian Ecology Center, 1983.

Byron, William. *Toward Stewardship: an Interim Ethic of Poverty, Power, and Pollution.* New York: Paulist, 1975.

Carmody, John. *Ecology And Religion: Toward a New Christian Theology of Nature.* New York: Paulist Press, 1984.

Cesaretti, C.A., and Stephen Commins. *Let the Earth Bless the Lord: A Christian Perspective on Land Use.* New York: The Seabury Press, 1981.

Chief Executives of Natural Resources Defense Council, et al. *An Environmental Agenda for the Future*. Washington, DC: Island Press, 1985.

Clark, W.C. and R.E. Munn, *Sustainable Development in the Biosphere*. Cambridge University Press, 1987.

Cobb, John B., Jr. *Process Theology as Political Theology*. Westminster 1982.

Corson-Finnerty, Adam Daniel. *World Citizen: Action For Global Justice*. Maryknoll, New York: Orbis Books, 1982.

Daly, Herman E. *Economics, Ecology, Ethics*. San Francisco: W. H. Freeman and Company, 1980.

Devall, Bill and George Sessions. *Deep Ecology*. Salt Lake City: Peregine Smith, 1985.

Dubos, René. *A God Within*. New York: Charles Scribner's Sons, 1972.

Eckholm, Erik. *Losing Ground*. A Worldwatch Institute Book. New York: W.W. Norton, 1976.

Freeman, John A. *Survival Gardening*. Rt.4, Box 232, Brevard, North Carolina: John's Press, 1984.

Fritsch, Albert J. *Environmental Ethics: Choices for Concerned Citizens*. Garden City, New York: Doubleday Press, 1980.

_____. *The Contrasumers: A Citizen's Guide to Resource Conservation*. New York: Praeger Publishers, 1974.

_____. *Renew the Face of the Earth*. Chicago: Loyola University Press, 1987.

Foreman, Dave. *Ecodefense*. Tucson, Arizona: Earth First Books, 1985.

Foster, Richard J. *Freedom of Simplicity*. New York: Harper & Row Publishers, 1981.

Freudenberger, C. Dean. *Food for Tomorrow?* Philadelphia: Augsburg, 1984.

Fukuoka, Mansanobu. *The One Straw Revolution: An Introduction to Natural Farming*. Emmaus, Pennsylvania: Rodale Press, 1978.

Granberg-Michaelson, Wesley. *A Worldly Spirituality: A Call to Take Care of the Earth*. New York: Harper & Row, 1984;

_____. *Ecology And Life: Accepting Our Environmental Responsibility.* Word Books, 1988.

Gray, Elizabeth Dodson. "Changing Images of Woman, Man, and God." *Conference Synod of the Northwest, Presbyterian Church USA.* Stony Point Conference Center, November 8-9, 1985.

_____. *Green Paradise Lost.* Wellesley, Massachusetts: Roundtable Press, 1979.

Griffen, Susan. *Woman and Nature: The Roaring Inside Her.* New York: Harper & Row 1978.

Goudie, Andrew. *The Human Impact: Man's Role in Environmental Change.* Cambridge, Massachusetts: MIT Press, 1982.

Gutierrez, Gustavo. *Theology of Liberation.* Maryknoll, New York: Orbis Books, 1973.

Hall, Douglas John. *The Steward: A Biblical Symbol Come of Age.* New York: Friendship, 1982.

_____. *Christian Mission: The Stewardship of Life in the Kingdom of Death.* New York; Friendship Press, 1984.

Hall, Douglas John. *Imaging God: Dominion as Stewardship.* New York: Friendship Press, 1986.

Hart, John. *The Spirit of the Earth – A Theology of the Land.* New York: Paulist, 1984.

Hoffman, Douglas R., ed. *The Energy-Efficient Church.* Total Environmental Action, Inc., New York: Pilgrim Press, 1979.

Jantzen, Grace. *God's World, God's Body.* Westminster, 1984.

Jegen, Mary Evelyn and Manno, Bruno V., eds. *The Earth Is the Lord's.* New York: Paulist, 1978.

Lappé, Frances Moore and Joseph Collins. *World Hunger: Twelve Myths.* New York: Grove Press, 1986.

League of Women Voters, *Household Hazards: A Guide to Detoxifying Your Home.* Albany, New York: League of Women Voters, 1990.

Limburg, James. *A Good Land.* Augsburg, Minneapolis Publishing House, 1981.

Longacre, Doris J. *Living More With Less.* Scottsdale, PA, Herald Press, 1980.

Lutz, Charles P., ed., *Farming the Lord's Land.* Augsburg, Minneapolis Publishing House, 1980.

Marchant, Carolyn. *The Death of Nature: Women, Ecology, and the Scientific Revolution.:* San Francisco: Harper & Row, 1980.

Marshall, Joyce and Gene. *The Reign of Reality.* Dallas, Texas: Realistic Living Press, 1987.

McLuhan, T.C. *Touch The Earth: A Self-Portrait of Indian Existence.* New York: Simon and Schuster, 1971.

Meeker-Lowry, Susan. *Economics As If the Earth Really Mattered.* Santa Cruz, California: Catalyst. 1987.

Mesarovic, Mihajlo D. *Mankind at the Turning Point.* The Second Report to the Club of Rome. New York: Dutton, 1974.

Mische, Gerald and Patricia. *Toward a Human World Order.* New York: Paulist Press, 1977.

_____. "Bioregionalism and World Order," in *Breakthrough.* Spring/Summer 1985, pp. 6-9.

Mitchell, John. *The Earth Spirit: Its Ways, Shrines, And Mysteries.* New York: Crossroad Publishing Co., 1975.

Moltmann, Jurgen. *God In Creation: A New Technology of Creation and the Spirit of God,* New York: Harper & Row, 1985.

NACCE Proceedings 1987. "Christian Ecology: Building an Environmental Ethic for the Twenty-First Century." San Francisco: NACCE, 1988.

Naess, Arne. *Ecology, Community and Lifestyle: An Outline of Ecosophy*, Cambridge University Press, 1988.

Nelson-Pallmeyer, Jack. *Water: More Precious Than Oil.* Augsburg, 1982. (booklet),

Owens, Virginia Stern. *And The Trees Clap Their Hands: Faith Perspective and the New Physics.* Grand Rapids, Michigan: Wm. B. Eerdmans, 1983.

Pierce, Gregory F. *Activism That Makes Sense: Congregations and Community Organization.* New York: Paulist Press, 1982.

Porteus, Alvin C. *Preaching to Surburban Captives*, (Sermon "Our Stewardship of Earth: Ecology and Faith").Valley Forge, Pennsylvania: Judson Press, 1979.

Pritchard, Judith. "The Wholeness of Creation: An Exploration," in *God's Green World.* London: The Christian Ecology Group, 1983.

Regan, Tom, ed. *Earthbound: New Introductory Essays in Environmental Ethics.* Philadelphia: Temple University Press, 1984.

Resnikoff, Marvin. *Living Without Landfills.* New York: The Radioactive Waste Campaign, 1987.

Sale, Kirkpatrick. *Dwellers in the Land: A Bioregional Vision.* San Francisco: Sierra Books, 1985.

Schumacher, E.F. *Small Is Beautiful: Economics As If People Mattered.* New York: Harper & Row, 1975.

Secundo, Juan Luis. *Liberation Theology.* Trans. by John Drury. Maryknoll, New York: Orbis Press, 1976.

Shi, David. *In the Simple Life.* London: Oxford University Press, 1985.

Shrader-Frechette, K.S., ed. *Environmental Ethics.* Pacific Grove, California: The Boxwood Press, 1981.

Sider, Ronald J. *Rich Christians In an Age of Hunger: A Biblical Study.* Downers Grove, Illinois: Intervarsity Press, 1977.

Spretnak, Charlene, *The Spiritual Dimension of Green Politics.* Santa Fe, New Mexico: Bear & Co., 1986.

Squiers, Edwin R. *The Environmental Crisis: The Ethical Dilemma.* Mancelona, Michigan: The AuSable Trails Institute of Environmental Studies, 1980.

Swimme, Brian. *The Universe Is a Green Dragon: A Cosmic Creation Story.* Santa Fe, New Mexico: Bear & Co. 1984.

Taylor, Paul W. *Respect for Nature: A Theory of Environmental Ethics.* Princeton: Princeton University Press, 1986.

Taylor, Richard. *A Community of Stewards.* Augsburg, 1982. (booklet),

Teilhard de Chardin, Pierre. *On Suffering.* New York: Harper & Row, 1974.

_____. *Hymn of the Universe.* Trans. Simon Bartholomew. New York: Harper & Row, 1965.

_____. *The Future of Man.* Trans. Norman Denny. New York: Harper & Row, 1964.

The Earthworks Group. *50 Simple Things You Can Do to Save the Earth.* Berkeley, Earthworks Press, 1989.

_____. *50 Simple Things Kids Can Do to Save the Earth* and *The Next Step.* Kansas City, Andrews and McMeel, 1990.

Timberlake, Lloyd. *Africa in Crisis: the Causes, The Cures of Environmental Bankruptcy.* Philadelphia: New Society Publishers, 1985.

Tobias, Michael, Ed. *Deep Ecology.* San Diego: Avant Books, 1985.

Uhlein, Gabriele. *Meditations With Hildegard of Bingen.* Santa Fe: Bear & Company, Inc., 1982.

Valaskakis, Kimon. *The Conservor Society.* New York: Harper & Row, 1979.

Ward, Barbara and Dubo, Rene. *Only One Earth: the Care and Maintenance of a Small Planet.* New York: W.W. Norton, 1972.

Wilkinson, Loren. *Earthkeeping: Christian Stewardship of Natural Resources.* Grand Rapids, Michigan: Wm. B. Eerdmans, 1980.

Solar-Heated Churches

American Baptist

Church of the Valley
San Ramon, California

Covenant Baptist
Mesa, Arizona

First Baptist
Aberdeen, South Dakota

Mt. Zion Baptist
Norfolk, Virginia

Providence Baptist
Baltimore, Maryland

Assembly of God

New Life Assembly
Wakefield, Rhode Island

First Assembly of God
Woonsocket, Rhode Island

Church of God

Calvary Road
Carlisle, Pennsylvania

Church of the Brethren

Basset
Basset Virginia

Broadfording Fellowship
Hagerstown, Maryland

Cedar Creek
Citronelle, Alabama

Flower Hill
Gaithersburg, Maryland

West York
Thomasville, Pennsylvania

Episcopalian/Anglican

St. Barnabas in London
London, Ontario

St. Paul's
Brookline, Massachusetts

Evangelical Covenant

Brookwood
Kansas City, Kansas

Lutheran

Christ Evangelical
Santa Fe, New Mexico

Good Shepherd
Tuckerton, New Jersey

Holy Trinity
Newinton, New Hampshire

Hope
Milton, Wisconsin

Peace
Otis Orchard, Washington

Shepherd of the Valley
Lakeland, Minnesota

St. Peter's Evangelical
Lamberville, New Jersey

Nondenominational

Center of Hope
Westminister, Colorado

Creede Community
Creede Colorado

Faith Christian Center
Bedford, New Hampshire

Faith Christian Fellowship
Elliot, Maine

Presbyterian

Burke
Burke, Virginia

The Church of the Hills
Bellaire, Michigan

Christo Del Valle
Albuquerque, New Mexico

First Presbyterian
Washingtonville, New York

New Covenant
Richmond, Virginia

New Life Community
Albuquerque, New Mexico

St. Andrew
Albuquerque, New Mexico

Trinity
Denton, Texas

Kanata
Kanata, Ontario

Reorganized Church of Jesus Christ of the Latter Day Saints

Monte Vista
Monte Vista, Colorado

Bethany
Bethany, Missouri

Roman Catholic

Holy Trinity Monastery Chapel
St. David, Arizona

Holy Trinity
Sherman, Connecticut

St. Catherine
Humansville, Missouri

St. Edward the Confessor
Medfield, Massachusetts

St. John Newmann
Lubbock, Texas

St. Patrick
Wheatland, Wyoming

St. Vincent
Connell, Washington

Presentation of Our Lady
Denver, Colorado

Southern Baptist

Emmanuel
York, Pennsylvania

Unitarian-Universalist

Eno River Fellowship
Durham, North Carolina

United Church of Christ

Caroline Mission
St. Louis, Missouri

Mira Vista
El Cerrito, California

United Congregational
Torrington, Connecticut

United Methodist Church

Asbury
Albuquerque, New Mexico

Community
Columbia, Missouri

Delran
Delran, New Jersey

Medford
Medford, New Jersey

Mt. Blanchard
Mt. Blanchard, Ohio

Scott Memorial
Cadiz, Ohio

St. Paul Center
Springfield, Oregon

St. Paul's
Idaho Falls, Idaho

Excerpted from *Sunbeams and Sanctuaries: An Informal Study of Solar-Heated Church Buildings in the U.S.A. and Canada* by Rev. Roy A. Johnson, Box 719, New Windsor, MD 21776.

Land and Rural Issues

The following are some general resources on land and food issues available from the Association for Public Justice.

Land and Rural Issues
Association for Public Justice
Box 56348 Washington, D.C. 20011

Cassette Tapes:

From the *Fourth International Christian Political Conference*: "Food, Farming, and Foreign Policy,"August 14-16, 1980:

Theodore R. Malloch, "Food, Farming and Foreign Policy: Questions for the Policy Makers."

Frances Moore Lappé, "Democracy, Power, Land, and Food."

Jean Lloyd-Jones, "The Politics of Land Use."

Marty Strange, "Energy and Agriculture: Industrialization and Lost Opportunities."

James W. Skillen, "Land Rights, Stewardship, and Justice."

The Public Justice Report:

The APJ Education Fund's regular publication (10 issues/year).

The APJ Position Paper:

"Justice for the Land, Land for the Caring."

Background Studies:

Calvin DeWitt, "Stewardship of the Land."

Theodore R. Malloch "Public Policy and the Alcohol Fuel Binge."

Kenneth Piers, "Structural Problems in North American Food Production."

Patricia E. Govert, "State Restrictions on Coastal Wetland Use: Private Property Rights Versus Public Rights."

Uko Zylstra, "Ecological Aspects of Food Production."

Nicholas D. Giordano, "Energy Resources and Agricultural Production in Less Developed Countries: An Ecological Perspective."

Bibliographies:

An APJ Education Fund Bibliography of recent articles on the issues of land use and control, food and fuel competition, and the structure of agriculture.

Vischer, Jack and Zinkland, Daniel. *A Bibliographic Guide to Information in Agriculture and Food.* Nearly 100 pages of journals, books, organizations, media sources, state committees, and other sources..

Slide Program:

Theodore Malloch and Kenneth Heffner, "Energy Stewardship."

Basic Books on a Christian Approach to Political Action:

James W. Skillen. *Christians Organizing for Political Service*. Washington, D.C. Education Fund, 1980.

>A study guide based on the work of the Association for Public Justice.

James W. Skillen, editor, *Confessing Christ and Doing Politics,* Washington, D.C.: APJ Education Fund, 1982.

>Essays.

Hunger Resources

Bread for the World

802 Rhode Island Ave. NE, Washington, DC 20018
(ATTN: Publication Order)

WORSHIP AIDS

Faces of U.S. Hunger.

Worship service highlights domestic hunger in context of biblical faith. Suitable for liturgical and free worship. Companion bulletin insert, "Voices of Hunger."

Voices from the Quiet.

15 minute chancel drama appropriate for worship services or other church gatherings. Cast includes narrator, two actors and chorus. Addresses problems of taking hunger seriously and getting involved. Companion bulletin insert, "We are the Hungry."

Singing for Power.

Meditation and responsive reading emphasizes importance of singing in lives of individuals with particular emphasis on oppressed people.

Lazarus.

An oratorio based on Luke's Lazarus account. Musical highlights problem of hunger and poverty in dual contexts of biblical faith and modern world. Designed for church choirs, campus groups, community drama and choral groups. Easy to learn for an ecumenical, interracial community choir at Thanksgiving, World Food Day or Lent.

BOOKS AND OTHER RESOURCES

Bobo, Kimberly. *Lives Matter: A Handbook for Christian Organizing.*

How-to manual for Christians. Covers organizing groups, working in the church, public speaking, and media use.

Hunger in the Land of Plenty.

Seven-session study guide introducing participants to US hunger issues. Group discussions, simulations, and action.

Youth Retreat Resource Packet.

A set of materials for a "Right to Food Retreat" for youth or young adult groups. 10-hour experience seeks to create an understanding of constraints of those living in poverty. Examination of attitudes and beliefs about poor people, develops awareness of local and national resources and provides suggestions for action. Packet includes worship aids, recipes, scripts and other information.

World Hunger Education Service

1317 G St NW, Washington, DC 20005

Hunger: An Organizing Guide for Education and Advocacy.

Lists national state, regional and local organizations working on domestic and international hunger issues.

National Committee for World Food Day

1001 22nd NW Washington, DC 20437 (202) 653-2404

Resource list of materials available to help plan events and programs on World Food Day. Materials include Media Kit for Local Organizers, Videos and Films on world issues, songs and other aids.

Ecological Merit of Simple Lifestyle Practices

A = Preserves physical health
B = Sound economics
C = Balances ecology (curbs pollution)
D = Offers social development
E = Expresses solidarity with the poor
F = Builds community
G = Expresses corporate self-reliance
H = Expresses appreciation for nature and life
I = Shares global resources
J = Changes institutions
K = Offers personal growth
L = Tames technology

	A	B	C	D	E	F	G	H	I	J	K	L
FOOD												
Refrain from:												
junk foods	x	x	x		x		x		x	x	x	
prepared foods		x	x		x		x		x	x	x	
red meat	x	x	x					x	x		x	
soft drinks	x	x	x							x	x	
Patronize:												
farmers markets		x		x	x	x	x		x	x	x	
organically grown foods	x	x	x	x	x		x	x	x	x		x
local gardening	x	x	x	x	x	x	x	x	x	x	x	x
harvest naturally grown foods (nuts, herbs, berries, roots, etc.)	x	x	x	x	x	x	x	x	x	x	x	
use solar food dryer or cooker		x	x	x	x	x	x	x	x	x		x
curb alcohol & caffeine use	x	x	x	x	x	x	x	x	x	x	x	
use seasonal foods	x	x	x	x	x	x	x	x	x	x	x	x

	A	B	C	D	E	F	G	H	I	J	K	L
conserve cooking fuel		x	x		x		x		x	x		
boycott selected food products		x		x	x					x		
reduce sugar use												

HOUSING

	A	B	C	D	E	F	G	H	I	J	K	L
select sites near work place		x	x	x	x	x	x	x	x	x		
refurbish existing buildings		x	x	x	x	x	x	x	x	x		x
use native materials		x	x	x	x	x	x	x	x	x		x
adapt structure to needs (expand in summer or child-bearing years)		x	x	x	x	x	x	x	x	x		x
use solar energy	x	x	x	x	x	x	x	x	x	x		x
use wood for back-up heat		x	x						x	x		
insulate		x	x						x	x		
conserve hot water		x	x					x	x			
refrigerate wisely		x	x					x	x			
use less lighting		x	x					x	x			
eliminate useless gadgets		x	x					x	x			

CLOTHING AND PERSONAL ITEMS

	A	B	C	D	E	F	G	H	I	J	K	L
reduce new clothes purchases/trade clothes		x	x	x	x		x		x			
mend and maintain clothes		x	x	x	x		x		x			

	A	B	C	D	E	F	G	H	I	J	K	L
avoid endangered species products (furs, skins, etc.)								x		x		
select good footwear	x	x	x									
choose fabric wisely & for durability	x	x	x	x			x		x			
refrain from use of cosmetics	x	x	x								x	
recycle materials		x	x	x		x	x	x	x	x		x
conserve wood & paper products		x	x					x		x		x
curb tobacco & drug use	x	x	x	x			x	x		x		
recycle bottles & cans		x	x	x		x	x			x		x
reduce packaging		x	x	x		x	x			x		x

GARDENING & LAND CARE

	A	B	C	D	E	F	G	H	I	J	K	L
use solar greenhouse & cold frames		x	x	x	x	x	x		x		x	x
convert lawns to edible plants and trees		x	x	x			x		x	x		
compost organic wastes		x	x	x			x		x			
use biological & integrated pest control	x	x	x	x				x	x			x
test soil		x	x						x			
use gray water to irrigate		x	x						x			
select and cultivate multipurpose trees		x	x					x	x			
develop multipurpose parking & outbuildings		x	x					x	x			

Merit of Simple Lifestyle Practices

	A	B	C	D	E	F	G	H	I	J	K	L
garden intensively	x	x	x					x	x		x	x

TRANSPORTATION

	A	B	C	D	E	F	G	H	I	J	K	L
use car less (bike or walk)	x	x	x	x	x	x	x	x	x		x	
carpool		x	x	x	x	x	x		x			
use public transportation		x	x	x	x	x	x		x			
use special transportation for infirm & elderly	x	x	x	x	x	x	x		x			
maintain car properly		x	x						x			
buy energy-efficient cars		x	x						x			x
drive at moderate speeds	x	x	x	x					x			
travel less		x	x		x				x			
dispose of oil properly		x	x			x			x			

RECREATION

Decrease:

	A	B	C	D	E	F	G	H	I	J	K	L
use of motorboats, snowmobiles, motorized campers, dune buggies		x	x	x					x	x		x
hiking in restricted areas		x	x	x					x			
illegal hunting		x	x	x					x			
TV, radio, VCR, use	x	x	x	x		x			x			

Increase:

	A	B	C	D	E	F	G	H	I	J	K	L
walking, hiking, biking, rowing, climbing, caving, cross-country skiing	x	x	x	x				x	x		x	

	A	B	C	D	E	F	G	H	I	J	K	L
participative sports	x	x	x	x				x	x		x	
reading			x	x				x			x	
non-competitive entertainment				x		x					x	x
constructive hobbies		x	x	x				x			x	
sports near home		x	x	x						x		

HEALTH & PERSONAL SERVICES

	A	B	C	D	E	F	G	H	I	J	K	L
provide quiet spaces	x		x	x		x		x			x	x
participate in physical fitness programs	x	x	x	x			x		x		x	x
practice natural childbirth & birth control	x	x		x	x	x	x	x		x	x	x
give home care to elderly, ill, & disabled	x	x		x						x	x	
practice good hygiene	x	x									x	
go for medical, dental, & optical check-ups	x	x									x	
supervise children's TV	x	x		x		x					x	
celebrate simple weddings, anniversaries, & funerals		x		x							x	

COMMUNITY SERVICE

	A	B	C	D	E	F	G	H	I	J	K	L
monitor & protect water	x	x	x	x		x	x	x	x	x		x
dispose of paper waste & use compost toilets	x	x	x	x		x	x	x	x	x		x

	A	B	C	D	E	F	G	H	I	J	K	L
recycle garbage			x	x					x	x		x
offer adequate fire, police, & health services	x	x	x	x		x					x	
legislate noise ordinances	x		x	x		x	x	x			x	
legislate land use ordinances		x	x	x		x		x				
reclaim vacant lots	x	x	x			x		x			x	
improve parks	x	x	x			x	x	x			x	
provide proper educational facilities				x		x						
provide proper medical facilities	x	x		x							x	
use solar energy for public events			x	x		x	x					x
protect wildlife				x		x	x	x	x		x	
control air pollution	x	x	x	x	x		x	x		x		x
offer environmental education			x	x							x	

Caring for the Earth

This exciting and very original book of activities will help children of the 1990's see that a relationship exists between God and creation — and the activities show children what this relationship is.

LOVING OUR NEIGHBOR, THE EARTH
Creation-Spirituality Activities for 9-11 Year Olds
Christie L. Jenkins
Paperbound, $14.95, 125 pages,
8½" x 11", perforated
ISBN: 0-89390-204-7

Here are twenty easy-to-follow lesson plans that teach children (ages 9-11) how to recognize the value of all life. There are student handouts that can be photocopied and a variety of planned activities. A handy guide combines lessons into units easily taught in a Vacation Bible School, as a summer supplement, or along side a regular religious education program. Topics include water conservation, cultivating the earth, "God and garbage", and learning the stars.

Children of all faiths will grow in wisdom as their eyes open to the wonder of God's creation. When they finish these lessons they will understand what the author means when she says that polluting the earth is nothing short of sacrilege.

> *I found the book exciting... Dr. Jenkins suggests some valuable foundational steps that those of us involved in the ministry of children could begin to take....I hope that all teachers and catechists have the opportunity to see the suggestions with the issues proposed in* Loving Our Neighbor, the Earth.
>
> Mike Reifel
> Director of Catechetical Ministries
> Diocese of San Bernadino

> *The author is fully at home both in the theological and the ecological literature, and does a splendid job of bringing them in dialog with each other. Orientation for the teacher (which is theologically able and sensitive), creative lesson plans, and the aim of bringing action into teaching, make this a very fine work.*
>
> Dr. William A. Beardslee
> Professor of Religion, Emeritus
> Emory University

About the Author
Christie L. Jenkins has a doctorate in microbiology from the University of California and has done research at UC Davis and UC Riverside. She is currently an instructor in Biology at CSU/Northridge. She has also studied at the School of Theology at Claremont, California. Dr. Jenkins has a strong interest in linking faith and science for children and has taught on the undergraduate level. She has published in a variety of scientific journals.

Order through your local dealer, or use the order form on the back of this page.

MEDIA LIBRARY
Department of Religious Education
225 Elm Street
Youngstown, Ohio 44503

What is Geo-Justice? Geo-Justice is...

...a personal and planetary challenge, a new context for theological reflection, a way to discover our most timely task, the converging terrain between a spirituality of the Earth and social and environmental justice.

...a holistic and visionary call to aliveness, passionate and practical action for a deeper understanding of our mysterious journey in the universe that will open us to the beauty and crises of our time as we navigate through deep cultural waters while bringing balance to relationships and liberation to the oppressed earth.

...an invitation, in a moment of an exponentially growing sense of urgency, through rebirth and synthesis to a better next millennium through nothing less than a planetary Pentecost.

GEO-JUSTICE: Toward a Preferential Option for the Earth
Jim Conlon
Paperbound, $15.95, 200 pages, 6" x 9", ISBN 0-89390-182-2

What do you get when you combine an environmentalist with an advocate for the poor? A Christian. That's the good news as told in a groundbreaking new book by Jim Conlon, program director at the Institute for Creation-Centered Spirituality. **Geo-Justice: Toward a Preferential Option for the Earth** puts the fate of the earth into the context of the paschal mystery. He shows how the earth is dying, how it is being reborn, and how we are preparing for a planetary Pentecost that will bring us all together.

> "Jim Conlon contributes significantly to the unfolding of the sacred story by insisting that the Earth itself must be included among the oppressed. A timely book!"
> **Matthew Fox**, author of *Original Blessing: a Primer in Creation Spirituality* and *The Coming of the Cosmic Christ*

> "Jim Conlon helps bridge the often divided worlds of those struggling for social justice and those seeking a creation-centered spirituality."
> **Patricia Mische**, Co-Founder, Global Education Associates

About the Author
James A. Conlon is associate professor at Holy Names College in Oakland and Program Director of its Institute in Culture and Creation Spirituality. He has doctorates from Union Graduate School and Columbia Pacific University. He has been Director of Field Education for the Toronto School of Theology, and holds degrees in Theology and Chemistry.

ORDER FORM Order these books through your local dealer, or complete this order form and mail it to:

QTY   TITLE   PRICE   TOTAL

_____
_____
_____

Subtotal:_____

CA residents add 7¼% (Santa Clara Co. add 8¼%) sales tax:_____

*Postage and handling:_____

Total amount:_____

*Postage and handling:
$2.00 for orders up to $20.00
10% of orders over $20.00 but less than $150.00
$15.00 for orders of $150.00 or more

Resource Publications, Inc.
160 E. Virginia St., Suite 290
San Jose, CA 95112-5876
or call (408) 286-8505
fax (408) 287-8748

My check or money order is enclosed.
Charge my ❑ VISA ❑ MC Exp. date:_____
Card#_____-_____-_____-_____
Signature:_____
Name:_____
Institution:_____
Street:_____
City_____St_____Zip_____